So Much Wind?

STRUAN STEVENSON has served as a Conservative Member of the European Parliament for Scotland since 1999. He is Chairman of the European Parliament's Delegation for Relations with Iraq, Senior Vice President of the Fisheries Committee and President of the Climate Change, Biodiversity and Sustainable Development Intergroup. He was appointed during 2010 by the Organisation for Security and Cooperation Europe (OSCE) as a Personal Representative (Roving Ambassador) of the Chairman in Office (Kazakhstan) responsible for Ecology and Environment with a particular focus on Central Asia.

In 2004 he won the Templeton Foundation $50,000 literary prize for his essay 'Crying Forever', about the plight of victims of Stalin's nuclear tests in East Kazakhstan. He donated all of this money to hospitals in the area. He went on to publish a book, also entitled *Crying Forever*, which was launched at the UN headquarters in New York in 2008, with a Russian version published in 2009. These publications, together with other generous contributions, have to date raised more than $100,000, which has also been donated to the victims of the nuclear tests in Kazakhstan through the worldwide charity Mercy Corps.

In August 2012, Struan launched his second book, *Stalin's Legacy: The Soviet War on Nature*, published by Birlinn Limited (www.birlinn.co.uk), at the Edinburgh Book Festival.

So Much Wind?

The Myth of Green Energy

STRUAN STEVENSON

BIRLINN

First published in 2013 by
Birlinn Limited
West Newington House
10 Newington Road
Edinburgh
EH9 1QS

www.birlinn.co.uk

ISBN: 978 1 78027 113 2

British Library Cataloguing-in-Publication Data
A catalogue record for this book is available from the British Library

Typeset by Iolaire Typesetting, Newtonmore
Printed and bound by Clays Ltd, St Ives plc

Energy and persistence alter all things
Benjamin Franklin

Contents

List of Illustrations

Acknowledgements

My sincere thanks to Ben Acheson for giving up countless evenings, weekends and holidays to contribute valuable research and ideas for this book, as well as spending endless hours checking and correcting facts and suggesting improvements; his helpful insights and meticulous research have been an inspiration. Also my great thanks to my publisher, Hugh Andrew at Birlinn Limited, for his invaluable guidance and inspiration for this book.

Foreword

We are facing a man-made energy catastrophe. It is not the catastrophe of melting icebergs, stranded polar bears and disastrous floods. It is a catastrophe brought into being by the enslavement of a government to one of the greatest ideological alarms of our age: the manufactured panic of climate change. On the strength of this alarm the Scottish government has embarked on a policy that is not only destroying our natural landscape – equivalent to, as one environmentalist vividly described it, taking a knife to the face of a Rembrandt – but it has also the potential to bring energy shortage and huge economic cost. There is a price we are going to have to pay. And Struan Stevenson in this book sets it out in unsparing detail.

The damage now being inflicted by wind farms, turbines and giant pylons on some of the most beautiful natural landscapes in the world is on a colossal and unrelenting scale. It has been the cause of countless appeals and objections by communities across the land. It has brought widespread and vocal objection. It has filled the letters pages of newspapers and protests from environmental and heritage bodies. In my years as Executive Editor of *The Scotsman* I can recall no other issue that brought forth such deep-felt opposition and concern among readers. For these, Struan Stevenson has become an eloquent spokesman. Many will find comfort in this book – the comfort of shared outrage.

But it is the setting out of the economic case against the wholesale resort to renewable energy that will render this book of inestimable value for many, not just in Scotland, but wherever governments are developing and implementing policies for renewable energy. This is a powerfully informed and formidably researched analysis of the folly of one particular government's policy. Few can read it and not be moved to the same strength of feeling that the despoliation of our landscape has aroused.

The case, simply put, is this. The energy capacity generated by wind farms has to be matched by reliable back-up supply for those periods when the wind is either not blowing or blowing too hard. This requires investment in gas, coal or nuclear. But the Scottish government has made no plans for the back-up required. Even if it were to come to its senses today there is just not enough time to commission any new coal-based plants with carbon capture and storage capabilities before 2015. And even if the First Minister Alex Salmond were to undergo a sudden Damascene conversion to nuclear, he could not commission a replacement nuclear plant and have it operational before 2019 at the earliest. 'So to avoid widespread black-outs at the end of the decade,' Struan Stevenson points out, 'we will have to hope that large numbers of open-cycle gas turbines are built between 2012 and 2015. But even then, increased dependency on imported gas will increase demand. Increased demand will increase fuel bills and increased fuel bills will increase fuel poverty.'

To this subject, Struan Stevenson has brought the research, argument, courage and passion of a modern-day Emile Zola. *So Much Wind* deserves not only to be read by everyone concerned with energy policy, but also to be placed in every school and college alongside the flawed analysis of those whose ideological alarmism has led directly to this policy debacle.

Bill Jamieson
Executive Editor, *The Scotsman*

AUTHOR'S PREFACE

The Rape of Scotland

Chambers Dictionary defines the word 'rape' as meaning violation, despoiling or abuse. It is an evocative word, but well describes the scandal of industrial wind developments in our nation today. Wind turbines violate the principle of fairness by transferring vast amounts of money from the poor to the rich. They despoil our unique landscape and environment; they risk plunging the nation into a devastating energy crisis and, through noise, the flicker effect and vibration, they abuse the health and welfare of people and animals which have to live near them.

Former Tory Chancellor Nigel Lawson described the UK Government's policy of achieving 20% electricity generation from renewable sources by 2020 as 'a fatuous obsession'. Scotland's five times more ambitious targets are therefore worthy of some intensive scrutiny.

This book seeks to evaluate the Scottish Government's obsession with renewable energy, while at the same time looking at alternative sources of power that may prevent the lights going out across Scotland.

CHAPTER ONE

So Much Wind?

Over the next quarter century, global energy consumption is forecast to grow by 61%. Over 2 billion people still do not have access to any power at all. If oil at over $100 per barrel is painful for us, it is an impossible agony for developing countries. Some people predict that if we continue to rely on fossil fuels as our main energy source, we will exacerbate world poverty, face catastrophic increases in global temperatures, create freak weather conditions and cause sea levels to rise by over one metre, wiping out tens of millions of people worldwide.

Whether we accept these dramatic arguments or not, there is no doubt that we have to change. We cannot go on with our current addiction to high-carbon fossil fuels. We must conduct our lives doing as little harm to the world we live in as reasonable and practicable. Reducing CO_2 emissions in Europe by 20% by 2020 is a start, but we need to aim for zero CO_2 emissions and the technology is already here to achieve this goal.

The current oil, gas and coal technologies are in their twilight years. They are sunset technologies. As I will demonstrate later in this book, even nuclear power, which is an almost CO_2 emission-free energy provider, cannot supply the answer. It would require an estimated 200,000 new nuclear plants around the world to replace the base load energy currently provided by coal, oil and gas. The capital costs, security risks and unresolved question of nuclear waste storage render such a prospect obsolete. Of course nuclear power will play a role in any CO_2 emission-free future, but that role will be relatively small.

It is against this background that the Conservative/Lib-Dem Coalition Government at Westminster has agreed to meet the EU targets of a 20% cut in CO_2 emissions by 2020 by providing a generous menu of subsidies to encourage a greater share of energy from renewable sources. The Scottish Government has gone much further, setting a target of delivering the equivalent of 100% of the nation's electricity requirements from renewable sources by 2020. As this book was going to print, Scotland's First Minister, Alex Salmond, claimed that Scotland was well on its way to beating the ambitious targets and upped the ante by setting a new target of 50% of electricity from renewable sources by 2015. Salmond stated that he is 'aiming for a transformation – a re-industrialisation along the lines of a green economy' (European Energy Review 27/08/2012).

This ambitious target to cut CO_2 emissions and at the same time 're-industrialise' Scotland, mostly through the erection of giant wind turbines, has won plaudits around the globe for First Minister Alex Salmond, who says Scotland can be the 'Saudi Arabia of renewable energy'. But the policy has increasingly attracted critics who have pointed out that the Scottish Government's obsession with wind power is both ludicrously inefficient and hugely costly and that the derisory amount of power produced by wind turbines cannot significantly contribute to meeting our electricity needs. The wind may blow for free in Scotland, but harnessing it as a power source at the public's expense is nothing short of economic folly and may indeed cause our lights to go out one day.

The SNP-led Scottish Government was elected because the people of Scotland trusted them to create and administer public policy on their behalf. They were trusted to spend taxpayers' money wisely and they were trusted to provide a secure supply of energy to their electorate. This government has betrayed that trust. They have hijacked energy policy as a vehicle for political rather than economic gain. They have caused our electricity and heating bills to skyrocket and they have forced over a third of Scottish homes into crippling fuel poverty.

I am certain that they did not set out on this path maliciously, but having trumpeted their intention to achieve 100% electricity generation from renewables by 2020 in Scotland, they have found themselves on a political juggernaut that is hard to stop. Now, despite mounting public opposition and clear evidence that their obsession with wind power is unsustainable, they plough on regardless, determined to save political face at any cost.

They are currently attempting completely to replace fossil fuels and nuclear power with renewable energy. They seem not to realise that commercially available renewable technologies simply cannot provide energy security. Giant industrial wind turbines are visual monstrosities that produce a trickle of electricity at vast cost to the consumer. They don't significantly reduce CO_2 emissions and they threaten to plunge us into an energy crisis, just as we try to drag ourselves out of a financial one.

We know that every single megawatt of generating capacity from wind farms has to be matched by a reliable, affordable back-up supply for when the wind is either not blowing or blowing too hard. The proven, currently available technologies are gas, coal or nuclear. But nobody from the Scottish Government is planning for coal or gas and they have made their position quite clear about nuclear.

Even if SNP Government ministers woke up tomorrow realising the error of their ways, there is simply not enough time to commission any new coal-based plants with carbon capture and storage capabilities before 2015. Even if he wanted to, Alex Salmond could not commission a replacement nuclear plant and have it fully operational before 2019 at the earliest. So to avoid widespread blackouts at the end of the decade, we will have to hope that large numbers of open-cycle gas turbines are built between 2012 and 2015. But even then, increased dependency on imported gas will increase demand. Increased demand will increase fuel bills and increased fuel bills will increase fuel poverty.

The Scottish Government has either failed to realise this or they have completely ignored it. They continue on their relentless stampede for renewables, which has become their flagship policy, second only to independence in order of importance. They aim to achieve their target of 30 GW of installed capacity by 2020, which would mean virtually doubling every year the number of on- and offshore wind farms currently under construction.

Operational Wind Farms				
Onshore			Offshore	
England	126	992.98MW	13	1,698.20MW
Northern Ireland	34	440.89MW	–	–
Scotland	141	3,124.51MW	1	10.00MW
Wales	38	423.40MW	2	150.00MW
Total	339	4,981.78MW	16	1,858.20MW

Total operational wind farms: 355 (6,839.98MW)

Wind Farms Currently Under Construction				
Onshore			Offshore	
England	33	547.95MW	6	1,794.90
Northern Ireland	4	59.30MW	–	–
Scotland	30	1,306.92MW	–	–
Wales	5	111.00MW	1	576.00MW
Total	72	2,025.17MW	7	2,370.90MW

Total wind farms currently under construction: 79 (4,396.07MW)

Consented Projects				
Onshore			Offshore	
England	104	896.03MW	6	2,328.00MW
Northern Ireland	34	446.60MW	–	–
Scotland	117	2,108.35MW	1	7.00MW
Wales	15	449.23MW	–	–
Total	270	3,920.21MW	7	2,335.00MW

Total Consented Projects: 277 (6,255.21MW)

Projects in Planning				
Onshore			**Offshore**	
England	104	1,108.06MW	4	1,854.90MW
Northern Ireland	41	619.35MW	–	–
Scotland	180	4,659.09MW	3	1,450.00MW
Wales	37	1,204.33MW	–	–
Total	362	7,590.83MW	7	3,304.90MW

Total projects in planning: 369 (10,895.73MW)

(Tables taken from Renewable UK website – August 2012 http://www.bwea.com/statistics/)

But even the experts struggle to justify such unrealistic targets. The Adam Smith Institute and the Scientific Alliance recently noted that to deliver 18 or 19 GW of offshore wind, the Government would need to construct another 5,000 turbines before 2020. With only around 120 days per year suitable for offshore construction, that means five turbines will have to be installed every day until 2020. It is just not possible.

In defence of their energy policy the Scottish Government often cites Denmark as a 'success' story. Denmark has put massive investment into onshore and offshore wind power and wind has become a major component of their energy mix, but it still only comprises around 20% of all power generated, although they only ever use half of what is generated. Alex Salmond sees Scotland overtaking Denmark to become the 'green energy capital of Europe' (European Energy Review 27/08/2012). This obsession with wind power may lead to blackouts in the future and, in any case, Denmark simply cannot be compared to the UK or Scotland.

By virtue of its geographical position, Denmark lies in an 'electrical transit corridor' with Sweden and Norway to the north and Germany to the south. The benefit of this corridor is that when wind power in Denmark exceeds the limit that their grid can cope with, the power is simply sent to Scandinavia and

Germany. Conversely, when Danish wind farms are not working, there are net power inflows, mostly from Norwegian and Swedish hydropower, which balances the Danish grid system and prevents electricity chaos.

The amount of wind the UK experiences is similar to Denmark and our wind farms have similar load factors. However, geographical positioning is the all-important issue. The UK is not in an electrical transit corridor. It is part of an 'electricity island system'. Therefore, balancing the grid is much more difficult. In Scotland, if we generate too much wind energy, we risk overloading the grid, because of the limitations of the cross-border interconnector. A similar problem exists north of Beauly. That is why the National Grid has to pay millions to power companies to shut down their turbines when there is too much wind.

Grid instability is a serious problem. It is also a costly one. In Germany, sudden fluctuations in the power grid cause major damage at large industrial firms. This has led a few companies to invest in their own individual power generators and regulators in order to reduce the risk of costly damage. Many more executives are now considering freeing themselves completely from Germany's electricity grid in order to avoid the negative consequences of transitioning to renewable energy. The problem for the Germans is that their grid operators simply cannot predict how strongly the wind will blow (*Der Spiegel* 16/08/2012). Needless to say, Scottish grid operators will encounter the same problem.

There is also another glaring discrepancy in the Scottish Government's energy policy. Energy Minister Fergus Ewing recently said that subsidies paid to Scotland's wind farm operators and landowners come from consumers 'spread across the UK . . . since this is how the Renewable Obligation operates and will continue to do so'. So what if Scotland becomes an independent nation before 2020, fulfilling the SNP's other key objective – who will have to meet the costs of our renewables then? Does Fergus Ewing really expect us to believe that the English would be happy to continue footing the bill for Scotland's prohibitively expensive and ludicrously ambitious energy targets?

But challenging the new renewables religion is a high-risk strategy. There is a common perception, particularly amongst city dwellers, that wind turbines are clean and green. The usual myths are trotted out: 'They actually enhance the rural landscape and in any case beauty is in the eye of the beholder'. Such assertions are generally made by people who have only ever seen wind farms on the TV, or at a distance while speeding along the motorway in a car. These are the same people who believe that being opposed to wind turbines is the moral equivalent of climate change denial or voting BNP! They are certainly not people who have to live next to a wind turbine the height of a 65-storey tower block and who have seen their health suffer and the value of their homes plummet as a result.

The city-dwelling wind farm champions describe those who oppose wind turbines as 'a vocal minority'. This is true inasmuch as it is a minority who inhabit the countryside nowadays and have to live next to these industrial monsters. And so we hear government ministers at Westminster and Holyrood constantly reminding us that repeated surveys show a majority of citizens favour wind energy. We can hardly be surprised that any urban survey of people who never see a turbine will reveal that they are in favour of its use. But once they begin to understand the truth about the inefficiency of these turbines, their disastrous impact on the countryside and the billions in subsidies they absorb, coming directly out of the pockets of consumers, then perhaps in years to come the chattering classes may cease their chatter and change their views.

Rising Demand

Member states of the International Energy Agency (IEA) (including the UK, USA, Australia, Japan and 22 other countries) consume 45% less energy than they did in 1973 to produce the same unit of economic output. Energy consumption per capita has fallen by about 30% in the same period. However, absolute demand for energy continues to rise. Consumption by EU member states is predicted to increase by 30%, that of the USA by 42%, and that of developing countries by around 130%. By 2030, the predicted increase for China is 119% and for South Asia (including India) 188%.

This is happening in the context of energy reserves being finite and today the collective dependence levels of the 27 EU member states on imports are 77% for oil, 51% for gas and 34% for coal and solid fuels. By 2030, the overall dependence will have risen to 70%, with the figures for oil and gas expected to rise to around 80%, and for coal and other solid fuels to some 65%.

Europe needs a coherent energy policy. We cannot continue with 27 different member states pursuing 27 different and diverse energy policies working in isolation. We need strategic stocks for use in emergencies. We need member states to co-ordinate energy policy in a cohesive way. We need an open, competitive, single market in energy. Right now this just doesn't exist. Even the European Commission has been accused of driving the rush for renewables in an improper manner.

We have seen the increasing way in which Russia has used its stranglehold on oil and gas supplies to the EU. In 2003 they halted shipments of oil through the inappropriately named 'Friendship' or *Druzhba* pipeline to Latvia. In 2006 they stopped supplying Lithuania through this same pipeline. Also in 2006, Russia shut down its gas supply to the Ukraine for the first time in 40 years.

The problem is that at present we rely on Russia for 32% of our oil and 50% of our gas. As North Sea oil and gas reserves decline, our dependency on Russian imports will increase. This places the EU in a precarious position and, as relations with Vladimir Putin continue to deteriorate, we are increasingly at risk of being mauled by the Russian bear. We need to be less dependent on Russian energy and that means making the break from dependency on imported oil and gas. We also need to reduce EU dependency on oil imports from areas of great political instability such as the Middle East.

The other reason why energy has become a political hot potato is climate change. The increasing energy requirements of an expanding and industrialising global population, burning wood, oil, coal and gas, are leading to a steady rise in the amount of carbon dioxide released into the atmosphere. Whereas 200 years ago the earth's atmosphere contained around 280 parts per million (ppm) of CO_2, in 1999 the CO_2 level had reached 367 ppm. On current predictions of global energy use, that level will rise to 440 ppm by 2020 – almost as great a rise in the next 20 years as in the previous 200.

But these changes in the chemical balance of the atmosphere are of more than academic interest. Because an increased proportion of CO_2 in the atmosphere holds in more of the sun's heat, it raises global air temperatures. In turn, higher air temperatures will affect global climate (stronger winds, more rain) and cause sea levels to rise, perhaps by up to one metre or more. In countries like Bangladesh or, nearer to home, The Netherlands, the effects may be catastrophic. Tackling climate change cannot be the unique pursuit of an individual nation. It must be a global effort embracing the latest technological advances, combined with an international drive for efficient energy use.

Carbon dioxide is emitted as a result of the burning of fossil fuels. We therefore need to reduce CO_2 emissions dramatically. The fight against global warming has become a key priority. The USA, despite not ratifying the Kyoto protocol, has set up a well-targeted policy to reduce CO_2 emissions. At state level some initiatives have been developed on voluntary emission caps and regional carbon markets. This opens up the possibility of finding future common ground on concrete joint initiatives.

Some people advocate the use of 'clean coal' as a way of reducing CO_2 while providing a reliable source of energy production. However, 'clean coal' is so named because it reduces sulphur dioxide (SO_2) emissions by 90% and increases efficiency. Reductions in CO_2 are much less impressive, however, at around 25%. Although it can play a useful role in a future, diverse energy mix, 'clean coal' alone is not the answer.

It is for all of these reasons of energy security, combined with the global battle against climate change, that the Scottish Government has set its sights on renewables as the answer to all of our problems. Foolishly, they have ruled out the need for a more balanced approach. They are convinced that wind, wave, tidal and hydro power is the answer. They are dead set against allowing any new nuclear capacity to be built in Scotland and, apart from plans approved by ministers in 2011 to replace the 40-year-old coal-fired plant at Cockenzie in East Lothian with a gas-fired plant, there is little sign of preparations for any other gas or coal plants in the pipeline. Their obsession with renewables may lead to a catastrophic failure of our electricity supply system in Scotland.

CHAPTER TWO

Democracy Blowing in the Wind

Kay Siddell retired from a career as a civil servant to the rural tranquillity of South Ayrshire, where she and her husband John bought a smallholding near Dailly, a few miles east of Girvan, in what has been described locally as 'the secret valley'. Kay and John had hoped to settle down to the rural good life, away from the frenetic pace of the UK's cities.

They bought a beautiful property called High Tralorg Farm in 1988. Their view of the rolling Ayrshire hills was interrupted only by a single farmstead on the opposite side of the valley and a tiny, red-roofed cottage. Sadly, their dreams were shattered in 2006 when Scottish and Southern Energy plc (now SSE) was given planning approval to erect 52 giant, three-bladed Danish industrial turbines immediately facing their home on Hadyard Hill.

The diameter of the blades is 80 metres (262ft) and each turbine is mounted on a tubular steel tower. The wind farm includes three permanent 60-metre-high (197-ft) anemometer towers to monitor wind speeds, and is connected by a high-voltage overhead transmission line, which joins into the national electricity network at Maybole. The nearest turbine is only 742 metres from Kay and John's front door.

When I visited their home, I stopped at the top of an 800-metre rough farm track to open a gate that led to their farm. The huge turbines towered above me. I paused to wait for the noise of a high-flying jet to pass overhead so that I could listen for the sound of the turbines. Suddenly I realised that this was no high-flying jet. The noise was constant and was coming from the turbines themselves. Kay explained that the noise is relentless and maddening. She said the day I visited it was in fact relatively quiet. Her husband produced a decibel meter which registered 60 decibels, a level at which industrial workers would be offered ear protection.

Kay and John say that their peaceful valley has been wrecked, along with their health. Stress and depression have taken their toll. 'The world constantly churns with the movement of the blades,' says Kay, who is recovering from breast cancer and suffers from breathlessness and acute anxiety.

And yet Kay and John were never offered any kind of compensation and the value of their property is now, in their own words, 'negligible'. They feel powerless against the onslaught on their home and health by big business and landowners desperate to cash in on the wind farm windfall. They cannot sell their property and they are so ill that they have stopped doing any maintenance on or improvements to it. They are forced to live in one room of their house and they keep all the curtains constantly shut day and night to block out the view. The ever-present noise of the turbines even keeps them awake at night.

Kay says she often has to sit in the toilet at the rear of the cottage for hours, as it is the only part of the house where you can't see or hear the turbines. Now planning applications have been lodged to construct another 19 giant turbines on the other side of their home, which will result in them becoming completely encased and surrounded. Theirs is a real human tragedy and yet their rights have been simply ignored by the legislators, planners, landowners and energy companies.

At the time it was commissioned, Hadyard Hill was the UK's most powerful wind farm. It was visited by Alex Salmond himself, when he became Scotland's First Minister and was invited to open an extension to the wind farm. It was here, within a stone's throw of the Siddells' home in South Carrick, that he famously proclaimed that 'wind is free'.

The impact of this industrial invasion on their rural retreat has been far from free for Kay Siddell and her husband. The planners have purposely ignored the recommended 2km separation distance between turbines and residential dwellings. Kay now suffers from autoimmune problems which she blames on the stress of living next to the constant pounding of the giant turbines.

Kay's predicament is shared by everyone who has to live near a wind turbine. During the day, people who live next to wind farms are subjected to a continuous flicker effect when the sun is behind the rotating blades of the giant turbines. At night, they say the moon flicker and constant 'thump, thump, thump' of the turbines prevents them from sleeping. However, Kay says Scottish and Southern Energy, who pay thousands in 'community benefit' to the local village, have offered no compensation to her and her husband. Their lives have been blighted and their future compromised.

Sadly the Siddells are only one of many thousands of families in Scotland who have had their rights trampled underfoot by the windrush. Norman (Norrie) Gibson bought a former farmhouse and outbuildings at High Myres, on Myres Hill at Whitelee, near Eaglesham, around 30 years ago. This isolated property was the answer to a dream for Norrie and his wife. They had an uninterrupted view across the moor to the entire Firth of Clyde, with the spectacular backdrop of Ailsa Craig, Arran and the Mull of Kintyre in the distance.

High Myres is accessed up a long Forestry Commission road leading from Eaglesham. The Forestry Commission started buying land on the Whitelee plateau in the 1960s for planting great swathes of Sitka spruce. The very poor land at Whitelee was only capable of sustaining either Sitka spruce or Lodgepole pine. Reports prepared for the Forestry Commission in the early 1960s indicated that the plateau would be a very difficult area to plant trees, partly due to the constant high winds and the extremely high rainfall (nearly 1.8 metres per annum), causing significant drainage problems ('From Peat Bog to Conifer Forest: An Oral History of Whitelee, its Community and Landscape' by Ruth Tittensor, Packard Publishing 2009). Nevertheless, widespread tree planting took place at Whitelee and soon Norrie Gibson found his view of the Clyde coast obscured by large banks of conifers.

But it was the arrival of the energy giants, lured by the tales of high winds at Whitelee, that destroyed the Gibsons' tranquillity forever. First to set up shop was the National Engineering Laboratory (NEL) at an altitude of 330 metres, in an unplanted area of the Whitelee Forest immediately behind Norrie Gibson's home on Myres Hill. In 1986, several huge experimental turbines and aerials were erected only 300 metres from Norrie Gibson's door. He described the noise from the NEL turbines as 'like a cement mixer in the sky'. He told me when the winds are high on Myres Hill, NEL's aerials 'shriek and scream like banshees'.

The work undertaken by NEL soon bore fruit, with the arrival of ScottishPower Renewables and their plans for Europe's biggest wind farm. Now Norrie and his wife are surrounded by over 200 giant turbines. There is not a single window in their house from which you can't see a turbine. They are living in the centre of a gigantic electricity factory, their lives and home ruined. Norrie says that on many occasions, 16-tonne trucks bringing the massive steel towers up the Forestry Commission track to Whitelee get bogged down and stuck, blocking access to his property for days on end. He and his wife are alarmed at what might happen if a fire breaks out. Sometimes they are stuck in their house for up to four days because of these blockages.

Both Norrie and his wife suffer from severe ill health. Norrie is fighting cancer and the stress of living in the centre of Europe's biggest wind farm is not helping. His 300-year-old house, once a haven of peace and tranquillity, is now a virtual prison. The endless visual churning of the turbines, the constant noise and the flicker effect from every conceivable angle has taken its toll on the Gibsons' health. But throughout this saga, their objections and pleas have been ignored. Like others in similar situations, they say they have never been offered a penny of compensation. Their homes, health and happiness seem to be of little importance in the race to cash in on the wind power gold rush.

Norrie is a trained engineer. He has even worked at Hunterston nuclear power station on the nearby Ayrshire coast. But he has little sympathy for the wind turbines that now surround and loom over his home. He says that on several occasions he has witnessed huge blades breaking free from the wind turbines during storms and crashing to the ground, sometimes hundreds of metres away. He lives in terror that one could land directly on his roof. He described how sometimes he can be sitting quietly in his house and suddenly he hears a noise 'like an express train coming'. 'This is when it [the turbine] shuts down,' he explained. 'It feathers itself against the wind to stop and change direction. It's like an express train coming in. It gives you a fright sometimes.'

Norrie also described the unbearable flicker effect. 'Our bedroom is on this side,' he says, gesturing to one end of his home, 'and on a sunny morning it's just chop-chop-chop-chop-chop-chop-chop, with the light coming into the room.'

Kay Siddell and Norrie Gibson's cases are by no means unique. Local protests are routinely ignored. Even when council planners refuse consent for big industrial wind farm developments, the Scottish Government simply calls in the application and overturns the local council's decision. After years of protest an 18-turbine windfarm at Carscreugh Fell near Glenluce, one of the most beautiful corners of rural Wigtownshire, was approved by the Scottish Government, despite widespread local opposition and rejection of the plans at every stage of the planning process by Dumfries and Galloway Council. Councillors refused the application in April 2011 due to concerns about its landscape, visual and archaeological impact.

Undeterred, the developer, Spanish company Gamesa Energy UK, appealed to the Scottish Government against the council's rejection and, in a complete dismissal of local democratic decision making, ministers have now ruled that the proposals can proceed.

A reporter appointed by the government found the proposed 'energy park' would increase the amount of power generated by renewable resources and 'would tend not to overwhelm the skyline'. She also concluded the proposal was in 'broad overall accord with the relevant development plan'.

Scottish Natural Heritage and Historic Scotland had asked for a total of six turbines to be removed from the scheme. However, the reporter found that this would not provide an improvement of 'enough significance to warrant the resultant major drop in energy output'.

Dumfries and Galloway's Labour MP Russell Brown said the decision went directly against the wishes of local people. He wrote to First Minister Alex Salmond asking for the council's rejection to be upheld. 'Even supporters of wind energy are having their patience really tested by the number of wind turbines being erected in Dumfries and Galloway,' he said.

He said local concerns had been 'cast aside' by the government and claimed the region was taking an 'unfair share' of the nation's wind turbines. 'I have written to the First Minister to call on the government to uphold the council's rejection, because such blatant disregard for the concerns of local people cannot continue,' he concluded. Russell Brown has even started an online petition to garner support against any further proliferation of wind turbines in Dumfries and Galloway.

This sorry tale of local democracy trampled underfoot in the race for renewables in Scotland is, sadly, not an isolated one. In the obsessive pursuit of unrealistic renewable energy targets, the protests of tens of thousands of Scots are being routinely ignored. And unfortunately, as we will see, democracy is not the only victim of their chaotic energy policy. Kay Siddell and Norrie Gibson are just two of a multitude of victims of the Scottish Government's obsession with wind power.

And the failure adequately to consult the public doesn't stop there. It has now been shown that the European Commission failed in its obligation to collate data from every member state on their National Renewables Energy Action Plans, which is a legal requirement under Article 4 of the European Renewable Energy Directive (2009/28/EC). This failure led Avich & Kilchrenan Community Council, in Argyll and Bute, to make a formal submission of a case arguing that the failure to collate evidence constituted a breach of the Aarhus Convention. Aarhus grants the public rights regarding access to information, public participation and access to justice, in governmental decision-making processes on matters concerning local, national and trans-boundary environmental issues.

The community council's case sought to hold both the UK and Scottish Governments to account via the United Nations. A document lodged by Avich & Kilchrenan Community Council with the United Nations Economic Commission for Europe (UNECE), Aarhus Convention Compliance Team, was accepted as valid for consideration and the case was upheld, striking a blow against the flawed consultation process at the heart of the government's renewables programme.

The Community Council convinced UNECE that emissions savings attributed to wind energy by the UK and Scottish Governments were false and that both the EU and the UK have systematically made claims which are neither transparent nor valid. This means that planning approvals for wind farms under the Electricity Act are invalid, since they have proceeded on an entirely false prospectus.

A final ruling on the case concluded that the EU and UK are in breach of the Aarhus Convention and that the EU has bypassed the vital democratic accountability which the Convention requires. In these circumstances, the EU will now be ordered to implement measures that will ensure the public is provided with all the necessary information and the opportunity to participate in the decision-making process at every stage. Only by ensuring that all options are open, and effective public participation can take place, can the conditions of the Aarhus Convention be met. This was a significant victory for the anti-wind lobby and demonstrated clearly their ability to hold the UK and Scottish Governments – and even the EU – to account. This David and Goliath victory may, however, have come too late. The Scottish Government's obsession with wind may now be too advanced for the brakes to be applied.

CHAPTER THREE

The New Clearances

Families were driven from the Highlands to make way for sheep during the infamous Highland Clearances. Now the threat of a rural landscape devoid of people has arisen again. But this time it's not the lure of profits from mutton and wool that threatens our unique Scottish landscape. It is the Scottish Government's obsession with wind power and the steady march of giant wind turbines across our hills and glens that will drive away the tourists and leave a dwindling rural population struggling to survive.

It seems that nowhere is safe from the new wind-farm orthodoxy. In late June 2012 the Scottish Government wrote to the Convention of Scottish Local Authorities (COSLA) asking them to earmark even more land for wind farm developments. A letter signed by two ministers – Fergus Ewing (Energy) and Derek Mackay (Local Government) – called on COSLA to ensure that areas of land identified as suitable for wind farms were to be included in all councils' future development plans. As a sop to councils who are struggling to deal with a deluge of planning applications for industrial wind turbines, the ministers said that the Scottish Government would provide an extra £300,000 to help planning departments cope. This ministerial instruction to COSLA will act like an open invitation to energy companies, landowners and farmers to join the race for wind. Local authorities that turn down applications for wind farms based on local knowledge and in response to local protests will find their decisions being called in by the Scottish Government and overturned.

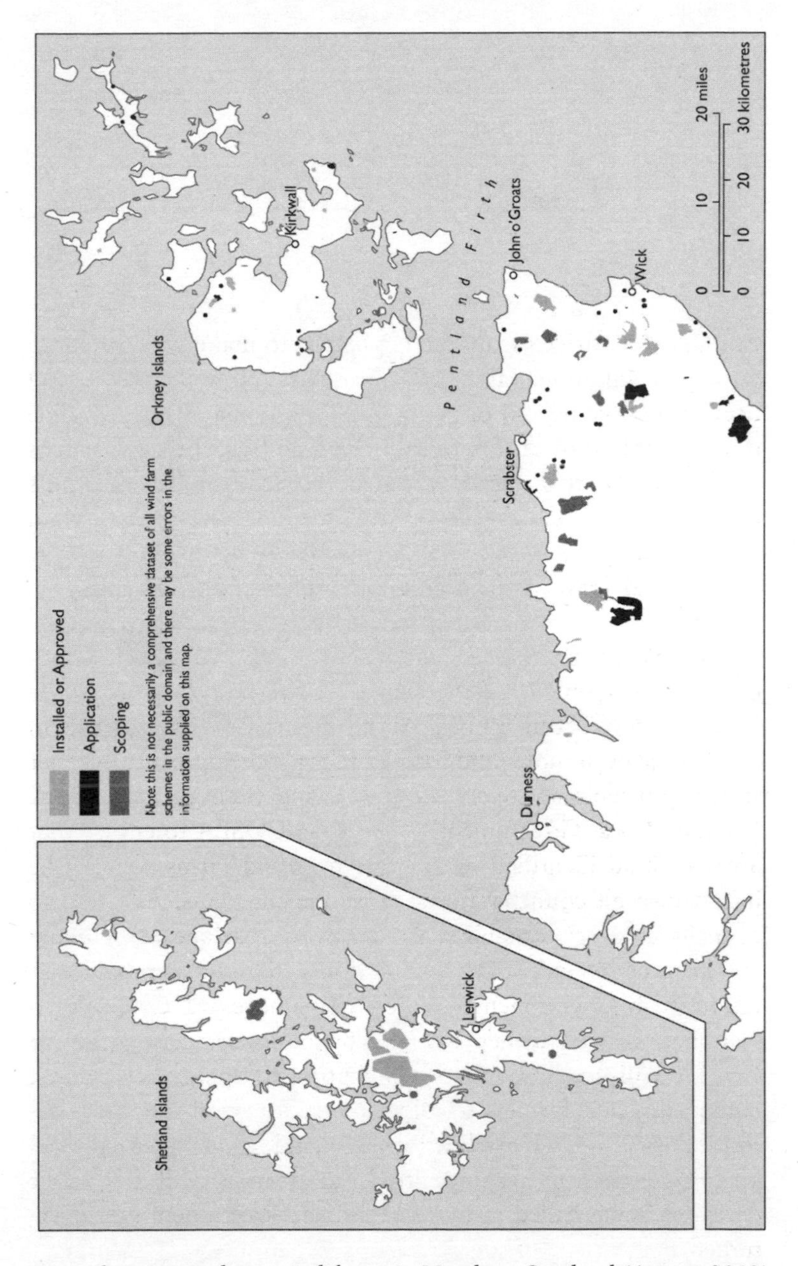

Map showing onshore wind farms in Northern Scotland (August 2012)

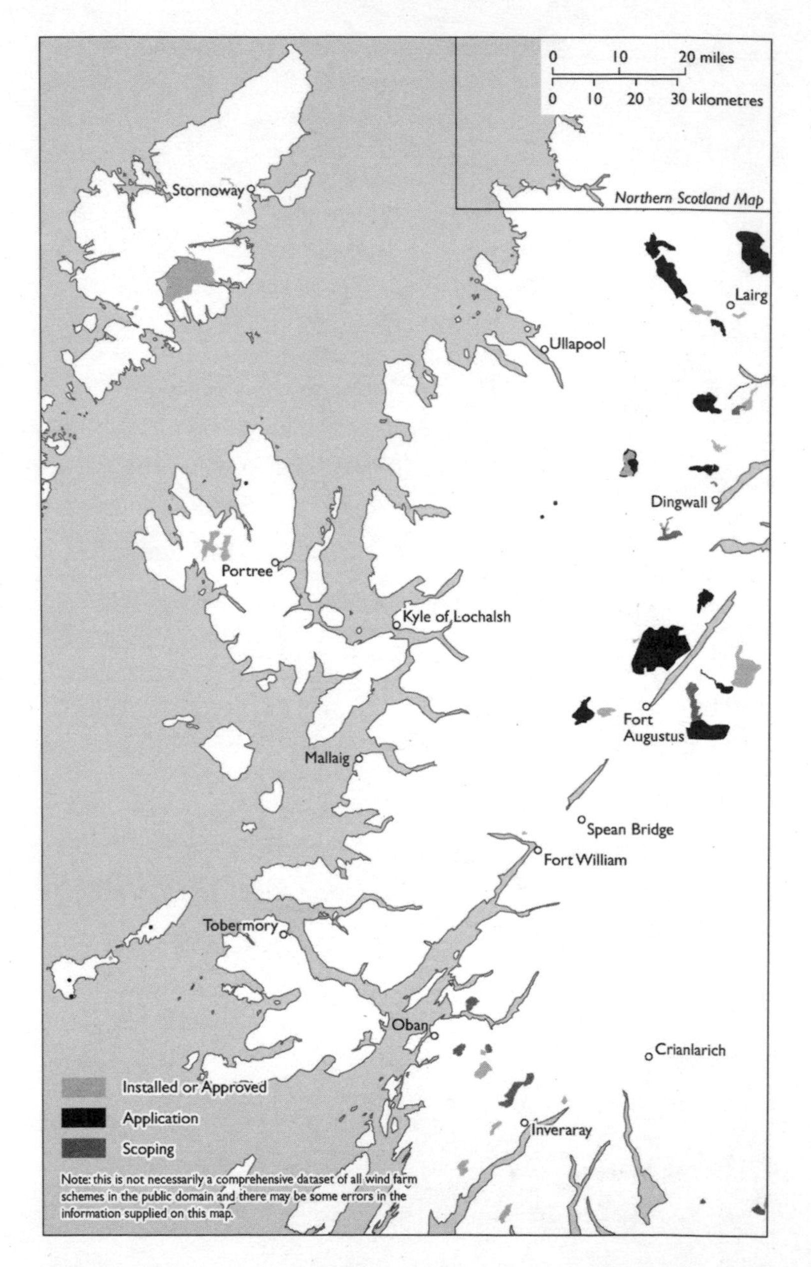

Map showing onshore wind farms in Western Scotland (August 2012)

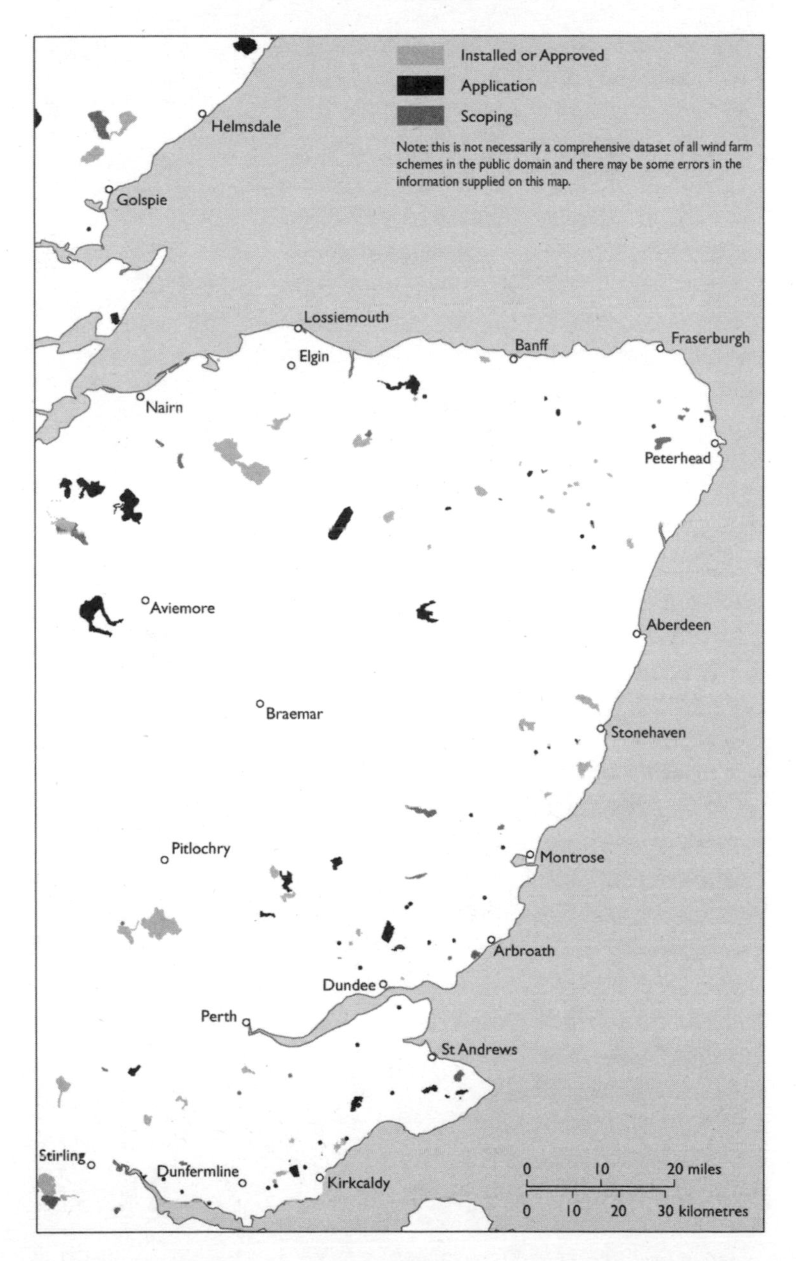

Map showing onshore wind farms in Eastern Scotland (August 2012)

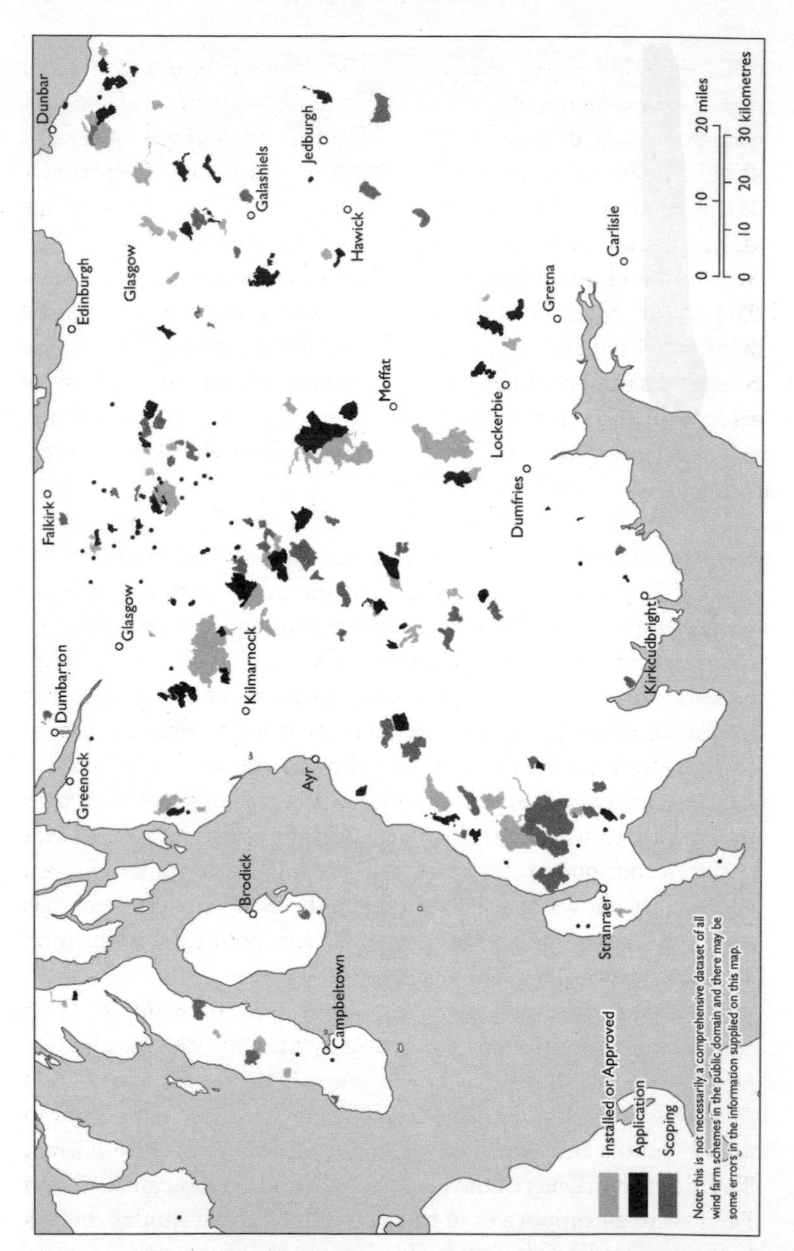

Map showing onshore wind farms in Southern Scotland (August 2012)

In September 2012, the Scottish Government announced that it will begin a major review of the country's planning policy to produce a third version of its National Planning Framework (NPF). In an address to the Scottish Parliament on 18 September 2012, Planning Minister Derek Mackay stressed the importance of the planning system in creating sustainable economic growth in Scotland and noted that NPF3 is due to be published by 25 June 2014. Mr Mackay stated that he wants the third National Planning Framework 'to reflect Scotland's ambition, to realise Scotland's opportunities for economic growth and to create more successful and sustainable places. This will contribute to supporting Scotland's economic recovery, particularly through the transition to a low-carbon economy.'

Put simply, the SNP Government has been facing increasing pressure over ill-conceived wind farm applications in unsuitable areas. They have realised that the spread of turbines will no longer continue unless they tailor the planning laws to suit their own political aims. Consequently, they want to introduce NPF3 in order to circumvent anti-wind protests and continue the unbridled development of wind energy. It is no coincidence that the timetable for NPF3 overlaps with the run-up to the Scottish independence referendum in autumn 2014. Is this merely a ploy by the SNP to use planning reforms for their own benefit, so that they can continue to claim that they are the 'world leader in green energy'? It shows that, once again, the SNP Government has hijacked energy policy as a vehicle for political rather than economic gain.

The democratic deficit doesn't apply only to wind farms in Scotland. The controversial decision by the Scottish Government to approve the Beauly–Denny power line, despite more than 18,000 letters of objection and only 46 letters of support, ran a knife across the throat of democracy and public consultation. The fact that Friends of the Earth, WWF and the (so-called) Green Party were all supporters of this scandalous act of vandalism lays bare their credentials as 'so-called' saviours of the environment.

It is also unbelievable that supporters of renewables seek to justify this environmental catastrophe with weasel words about Scotland becoming a European leader in clean, green energy. There is nothing clean or green about marching huge steel pylons across mountain landscapes and past some of Scotland's most historic castles and battlefields. The left-Green lobby spread the myth that turbines will end our dependency on fossil fuels. Nothing could be further from the truth. In fact, wind farms commit us to a fossil-fuel future due to the need for constant base load back-up from gas-fired or coal-fired power plants, which have to be kept permanently ticking over so that they can kick in when the wind drops.

The Scottish Government proclaims that exploiting Scotland's renewables potential is a national priority of the highest order and yet they have no coherent national plan, no programme or landscape capacity assessment of any sort and there is no national database or plan of what is happening on the ground, other than generalised figures of consented and installed capacity.

Nor is there any objective assessment and analysis of the actual environmental and economic benefits or effects, including the claimed benefits or effects associated with Habitat Management Plans. Everything is based on generalised assumptions and those who question these assumptions are dismissed as heretics.

The Scottish Government has powerful allies. Connie Hedegaard, the EU's Commissioner for Climate Action, has stated that 'people should believe that wind power is very, very cheap'. It is disappointing that someone in such a position of influence can be so deliberately misleading or just plain misguided. Compared perhaps to other renewables such as solar, tidal, wave and ethanol, wind power may indeed be a cheaper option, but as we shall see in detail later in this book, it is still one of the most expensive and inefficient ways to produce electricity today, perhaps as much as 200% more expensive than the cheapest fossil-fuel options and even this may be a serious underestimate.

With its 20-20-20 policy, the EU has promised that by 2020 it will have cut carbon emissions by 20% below 1990 levels, while increasing the total reliance on renewables by 20%. However, the

UK has taken the greenhouse gas reductions target a step further, aiming for a 34% reduction relative to 1990 levels by 2020. In Scotland, the target is 42%. Both the UK and Scottish Governments aim for an 80% reduction by 2050. The Scottish targets have been called into question by many professional observers as the policy appears to have been based on a misguided belief that if enough renewable technologies are deployed quickly, greenhouse gas emissions will be reduced.

Even the Institution of Mechanical Engineers, which is not against wind energy in theory, has slated the Scottish Government's ill-conceived plans. They question Alex Salmond's own 2020 targets, stating in a policy booklet entitled 'Scottish Energy 2020' (2011) that 'the government does not appear at this time to have a coherent plan to support delivery of this large-scale engineering task over the next eight years. Indeed, there are currently no credible strategies, from a technical point of view, published by government and available in the public domain.'

To achieve these targets, the UK Carbon Trust estimates that the cost of expanding wind turbines to produce 40 GW of power, in order to provide 31% of electricity by 2020, could run as high as £75 billion. But the benefits in terms of tackling climate change would be pathetic – a reduction of just 86 megatons of CO_2 per year for 20 years. This would mean that by the year 2100, the UK will have postponed global warming by just over ten days!

The Global Warming Policy Foundation estimated that to meet the targets for renewable generation, the UK will require a wind capacity of 36 GW backed up by 13 GW of open-cycle gas plants, as well as huge investment to update the electricity infrastructure. However, in their 2012 report 'Why is Wind Power So Expensive?', they warned that continuing the current focus on industrial wind power will postpone a temperature rise by a mere 66 hours by the end of the century despite costing the UK £120 billion per year! Amazingly, the report also suggested that the same electricity demand could be met from 21.5 GW of combined-cycle gas plants with a capital cost of £13 billion.

Worse still, these estimates may be overly optimistic, because the unreliability of wind means that we will have to rely on fossil-fuel coal and gas stations for base load back-up, and they will add to CO_2 emissions. With five-times higher targets than the rest of the UK for 100% electricity generation from renewables by 2020, Scotland will be trapped in an unsustainable time bomb of rising costs, fuel poverty, ruined landscapes and lost jobs.

The construction of Whitelee, the gigantic wind farm south of Glasgow mentioned previously, opened the floodgates for a mad dash for onshore wind. The project covers 55 square kilometres of Whitelee Forest and includes a 90-kilometre network of service roads which have clinically dissected Eaglesham Moor. A further 47 kilometres of access roads are planned during the ongoing extension phases at Whitelee, which will extend this gigantic wind factory by more than 200 turbines, each around 110 metres high. Now anyone driving along the A77 from Ayr to Glasgow is distracted by an alien landscape of huge metal towers and swirling blades (at least when the wind is blowing). The peace and tranquillity of Eaglesham Moor has given way to an industrial hell.

When I visited the Whitelee Visitor Centre in August 2012, I was fascinated to see tiny solar-powered wind turbines on sale in the shop. They had been reduced from £19 to £12 and their attractive box advised purchasers that this was the world's tiniest renewable energy turbine, powered by a miniature solar energy panel that was guaranteed for three years. I paid my £12 and, needless to say, when I got home and fitted the various parts together, I discovered that the toy turbine didn't work. It was useless . . . a perfect metaphor for the entire turbine sector! I was, however, saddened to find that the shop didn't sell MUGS with pictures of turbines on them. I wonder why?

Nevertheless, I did have some fun at Whitelee, which may have caused some nightmares for ScottishPower Renewables staff ever since. I was having a look around the interactive displays in the Visitor Centre and found one which was aimed mainly at children. I stepped on a platform, typed my first name 'Struan' into a computer and then had my face photographed in the middle of a giant wind turbine. This image was then saved on the Whitelee database for the edification of future visitors!

Brainwashing our kids

But interactive displays at Whitelee are not the only methods currently used to convince the next generation of Scotland's children that wind power is the Holy Grail of electricity production. Now the green fanatics are even spreading their zealotry in Scotland's classrooms, brainwashing our children into believing that the planet is doomed unless we tackle climate change by turning to renewable energy and adopting a 'green' lifestyle. Education Scotland has produced a teaching resource on climate change which states that using green technologies can help to reduce our impact on the planet. The programme has been condemned by leading scientists like Professor Tony Trewavas, a member of the Royal Society of Edinburgh and an academic from Edinburgh University, who says 'children should not be over-fed with one particular view of this. It is far too complicated for that.'

His view is shared by Martin Livermore, who runs the UK Scientific Alliance. He says: 'This is a clear attempt to push just one particular view – a view which, I'm sure, is widely held within the teaching profession. But, to my mind, it is brainwashing our children. I would like to see our children get access to a range of views. It is effectively moral blackmail to encourage children to adopt a green lifestyle for the good of the planet.' It is time our children understood that humans are only accountable for around 4% of global CO_2 emissions and even if we met the government's 2020 targets, Scotland's overall impact would be a risible reduction of less than 0.02%, if their policy achieved a reduction at all. There are those who argue that the race for wind energy may actually lead to an increase in CO_2 emissions.

Updated advice from the Scottish Government has instructed councils to include green energy in school curriculum or after-school activities 'to provide a foundation for balanced decision-making in later life'. The same guidance note also urged renewable power companies to redouble their public relations campaigns to ensure that 'the intermittent power and visual impact of turbines are not portrayed as show-stoppers or roadblocks'. Fergus Ewing, the Energy Minister, said the guidance would ensure wind farm planning applications 'go more smoothly for everyone involved'. The document was produced as part of the Good Practice Wind Project, a joint initiative between the Scottish Government and the EU which is aimed at examining obstacles to wind farm development (*The Telegraph* 22/08/2012). SNP ministers were immediately accused of spreading propaganda to the classroom by encouraging teachers to brainwash our kids.

Whitelee may be Europe's biggest onshore wind farm so far, but the eco-vandals have only just started. The stunning landscape at Dava Moor on the edge of the Cairngorm National Park near Grantown on Spey is set to be ravaged by a series of wind farms. This will lead to the destruction of deep peatland and even, in one case, the clearing of a native Scots pine forest planted with government grants just over a decade ago.

Kergord Valley in Shetland looks likely to be similarly vandalised in the face of huge local opposition, as does the Isle of Lewis, with the go-ahead given to a series of wind farms. As many as 39 turbines, each 145 metres high, at Muaitheabhal, will be visible from the world-renowned stone circle at Callanish. The area is also home to some of Scotland's most vibrant populations of white-tailed and golden eagles, which will be placed at immediate risk.

Plans have also now been unveiled for a second giant wind farm at Eishken on Lewis. Uisenis Power, owned by multimillionaire leisure tycoon Nicholas Oppenheim, is to erect 30 huge turbines next to the 39 which are under construction. It is thought that Mr Oppenheim will sell the development for over £200 million once planning approval has been granted. The RSPB has already warned that the new development will pose a serious threat to golden eagles.

On 8 September 2012 it was announced that the Scottish Government had given the go-ahead for yet another £225 million wind farm, only one mile from Stornoway. The wind farm will apparently pump millions in 'community benefits' into the Western Isles and a trickle of electricity onto the National Grid, when a new interconnector to the mainland is eventually built at huge cost. The RSPB Scotland withdrew its objections to the fact that the wind farm could kill golden eagles, when the developers (French energy firm EDF and engineering firm AMEC) agreed to reduce the number of giant turbines from 42 to 36. So eagles can count?

Even more bizarrely, the local (SNP) MSP Alasdair Allan suggested that some of the millions of pounds in community windfalls, which the developers had promised to pay, could be used to offset fuel poverty on the islands. The fact that fuel poverty in the Western Isles and indeed across the whole of Scotland is directly attributable to the mad dash to cover our landscape and coastal waters with gigantic and almost entirely useless wind turbines seems to have escaped Mr Allan's notice. He clearly thinks that the generosity of the giant energy companies in doling out millions of pounds in to local communities has everything to do with altruism and nothing to do with the enormous subsidies they rake in from consumers, who have to watch their electricity bills rise inexorably, month after month.

A New Planning Paradigm

One of the most significant problems we face when addressing the issue of a wind farm site is the current system of assessment that exists in considering the potential impact such a development would have on the land and local community. The EU's Environmental Impact Assessment (EIA) Directive currently in place gives the developer the responsibility to conduct the assessment and then gives the local authorities the task of evaluating their findings. Given that the UK has a considerable number of scientific experts with the ability to provide the relevant information related to the potential impact, it seems incongruous that the onus to do so lies solely with the wind farm's developer.

After having invested a great deal of time and money into a potential project, the developer would undoubtedly employ an expert who will advise in favour of the wind farm's construction. The EIA is, after all, currently used as a tool for persuading the planners and public to accept and approve the development. The local authorities themselves presently lack the essential scientific expertise to determine any potential project as damaging or counter-productive. Consequently, the local authorities rely on the relevant official bodies to make specialist observations about the proposed location on their behalf. These official bodies handle vast numbers of wind farm cases, bringing into question their ability to manage a case quickly, accurately and without an agenda. This is particularly the case when EIAs are of such magnitude that it would take council officials, never mind ordinary members of the public, weeks to read. The EIA for the proposed Moray Offshore Renewables Ltd (MORL) offshore wind farm in the Moray Firth was so vast, at over 70,000 pages, that it had to be delivered on a pallet by a forklift truck!

If there are three or four pages tucked away in the depths of this document that outline the damage to the marine ecosystem likely to arise from the installation of wind turbines on the seabed, it is improbable that many people will ever come across them. At three minutes' reading time per page, it would take a council official around 437 eight-hour days to read the entire report. One must assume that either the renewable energy company who ordered the report is being over-zealous in its attention to detail, or they are deliberately trying to bury information in a mountain of paper.

Wind farm developers have also become experts in sleight of hand. They produce computer-generated images of proposed turbines that make them seem much smaller than they are in reality. A study by Stirling University found serious flaws in the images that are routinely presented to councils and planners as part of the visual impact assessment process.

This prompted Highland Council in 2010 to establish their own standards, based on the Stirling University report, which provided the Highland public, planners and decision-makers with much more realistic and comprehensive visualisations. Needless to say, Highland Council's robust approach was fiercely resisted by the wind farm industry and their consultants, who have complained to Scotland's Energy Minister Fergus Ewing. Ewing has in turn asked SNH to review their 2006 guidance to local authorities in an attempt to find common ground. SNH, the landscape profession and the wind industry are now trying to steer through another solution to visualisation procedures, which, according to Alan Macdonald of Architech Animation Studios, the author of *Windfarm Visualisation – Perspective or Perception?*, is seriously flawed.

Proving that a site is unsuitable would have great financial implications. This is why overturning a decision to construct wind farms proves almost impossible for the local communities involved. This unfair balance built into the EIA process must be addressed. Local communities cannot compete financially with developers. This is particularly so with local rural communities, the very places where developers want to construct wind farms. The power currently lies with city investors, big developers and the local authorities, the very people who won't be directly affected by the proposals.

An EIA carried out by a local community would cost vast amounts of money and time including several days' field work, collating the field data, reviewing the EIA information, writing the report and even appearing as an expert witness at a public inquiry. The process requires one to consider a site's soil structure, archaeology, water quality, noise and ornithology, adding to the cost and time of an EIA.

So what is the solution? Local communities should be given a stronger voice in the planning process of any development that would significantly affect their local area. A fund should be provided which can be used by the local community to create their own EIA if necessary. This would guarantee further examination of assessments already carried out by local developers to ensure a full independent investigation of *any* development, not just wind farm construction.

Developers of major projects that will have an impact on the environment have to submit a financial bond to the local authority, which is earmarked for the decommissioning of the wind farm at the end of its working life. A separate bond should also be set up and used by the local authority to employ autonomous experts to assess the EIA or indeed to prepare an entirely separate and independent EIA. This local community bond could be applied to the initial planning application and could be based on not only the physical scale of the development but also on the economic costs.

Not only would this bond provide the local community with the power to object to any proposal but it would encourage the developers to research thoroughly the implications of such proposals and to produce good EIA documents which local councils would be happy to accept without having to consult other official bodies. This would encourage a win-win situation whereby developers produce high-quality EIA documents and local communities feel content that the documents represent the true potential impacts of such a development. Such a process would also reduce the time-consuming, costly and conflicting system currently in place while at the same time assisting those who feel disenfranchised by the planning process.

The current system is deeply unfair and heavily biased against local protesters who wish to protect their communities. Even fees payable to councils by energy companies for planning applications to erect giant industrialised wind farms amount to no more than £16,000. These fees should immediately be raised to £100,000 to make the planning process more affordable for local councils. At the current level of fees, councils end up out of pocket and the extra costs of dealing with these hundreds of egregious planning applications get passed down to local council tax payers, who also have to find the cash privately to pay for any fight against the proposals. So they end up paying twice! In

England, current fees for planning applications can cost up to £250,000, meaning that Scotland has become the cheaper alternative for any would-be wind farm developers. This is unacceptable.

In the meantime, rather than looking at ways to carve up Scotland's landscape and seabed in pursuit of its obsession with wind power, the Scottish Government needs to undertake a radical overhaul of zoning that considers a full appraisal of the potential collateral damage that developments can cause. This is the approach already taken in Germany, where both onshore and offshore areas have been carefully zoned to show where wind farms can be built and where they can never be considered. In Germany, areas of great scenic beauty, historical monuments and sites of environmental significance are no-go areas. In Scotland, they are fair game for renewables zealots and speculators to propose more and more turbines.

CHAPTER FOUR

The Financial Scandal

According to the preliminary findings of a controversial report by KPMG entitled 'Rethinking the Unaffordable', the UK Government could save each member of the population almost £550 by 2020 if it abandoned its plans for expensive wind energy in favour of cheaper gas-fired and nuclear plants (KPMG, November 2011). The report stated that Britain could still achieve its target of 20% reduction in greenhouse gas emissions by 2020, as set by the EU, without recourse to unreliable and costly wind energy, saving £34 billion in the process. However, the leaked statistics caused such a media storm that KPMG has refused to publish the full findings (*The Guardian* 07/02/2012).

This sounds like a fairly convincing argument, so why does the UK press ahead with the race for renewables? Why does Britain have the most generous subsidies for renewable energy in Europe? Even the decision by UK Chancellor George Osborne in July 2012 to cut subsidies by 10% and review the position again in 2014, still leaves Britain at the top of the subsidy league. So how did it happen?

The term 'energy' is often confused by politicians and the media alike with 'electricity'. To be specific, 'energy' is expressed in joules and is the amount regarded as the actual quantity of energy supplied (or consumed). 'Electricity' is measured in watts and is normally referred to in terms of 'power'. The Institution of Mechanical Engineers has commented that in the energy sector, 'power' is often wrongly considered to equate to 'electricity' whilst 'energy' is frequently used to denote the non-electrical forms of energy like 'heat energy'. This may be the source of the confusion. But energy does not mean electricity alone. In the UK, energy consumption is often split between three main areas of demand: electricity, energy for transport and heat energy. In Scotland, electricity is projected to be the smallest component of overall energy demand. Heat and transport energy are greater.

Whenever observers make declarations about the 'power' or 'energy' produced by a wind farm, it is vital to distinguish the differences between 'installed capacity' and the actual amount of energy supplied or consumed. Installed capacity is measured in megawatts (MW – a million watts), gigawatts (GW – a billion watts) or terawatts (TW – a trillion watts). Installed capacity is defined as the maximum amount of output produced by a given plant.

On the other hand, the actual amount of energy or electricity produced is time variable and is therefore measured in hours, i.e. GWh (gigawatt hour) or TWh (terawatt hour). Therefore, it is vital to distinguish between 'dispatchable' and 'intermittent' forms of power. Dispatchable forms of generation are those like nuclear or gas, which can be activated (with varying start-up speeds) to match demand. For example, a conventional nuclear power plant is normally operational and generating at full capacity for more than 80% of its life.

Conversely, wind and other renewables are intermittent. The amount of power they can generate depends on the wind's velocity. If there is no wind, they will generate nothing. They can only reach full generating capacity at wind speeds of a velocity of around 15–25 metres/second (50 mph) as any stronger wind forces the turbines to be shut down. Thus, a wind farm's output varies depending on the wind speed rather than consumer demand. This makes it particularly difficult to forecast the actual output ahead of a given time.

The EU Renewable Energy Directive (2009) requires the UK to achieve a target of 15% of total energy demand from renewable resources by 2020. Overall Energy Policy in the UK is the remit of the Department of Energy and Climate Change, with only some aspects being devolved to the Scottish Government. Notably, the Scottish Government has control over the most effective mechanism for realising the SNP's manifesto commitment: planning powers for any power station with a capacity of 50MW or more. However, Scotland's only legally binding commitment is to achieve the same EU target for renewable energy supply as the rest of the UK – i.e. 15% by 2020. But the Scottish Government has decided to double this target figure by agreeing to 30% of total energy coming from renewable sources by 2020.

This ambitious goal will run alongside the even more ambitious figure of the equivalent of 100% electricity generation from renewables by the same date. These targets have not been backed up by a credible published plan, but current figures suggest that the projections for Scotland in 2020 will be total energy 183.1 TWh/y (1 Terawatt-hour per year = 1 billion KWh/yr), which can be split into: Heat 89.7 TWh/y (49%); Transport 55.0 TWh/y (30%) and Electricity 38.4 TWh/y (21%).

These figures illustrate firstly that the Scottish Government's energy policy focus on Scotland's electricity requirements is misplaced, when it is by far the smallest component of our energy needs and, secondly, that even if we were able to source 100% of these electricity requirements from renewables by 2020 in a reliable manner, this would barely achieve 20% of the energy target by 2020, never mind the hugely ambitious 30% now projected by Alex Salmond and his ministers.

With consumers paying for both the turbines and an updated electricity grid, you may be wondering why the UK and Scottish Governments are even entertaining such a policy? Well, it is because energy policy is inextricably tied up with climate change policy, which seeks to cut greenhouse gas emissions. Thus, low-carbon technologies and renewable technologies have been incentivised by an assortment of carrots and sticks.

In their 'Reforming Energy Subsidies' policy booklet, The United Nations Environment Programme (UNEP) noted how there is widespread confusion regarding what exactly an energy subsidy is. In its simplest form, the UNEP booklet defines it as 'a direct cash payment by a government to an energy producer or consumer to stimulate the production or use of a particular fuel or form of energy'. The Organisation for Economic Cooperation and Development (OECD) sees an energy subsidy as 'any measure that keeps prices for consumers below market levels, or for producers above market levels or that reduces costs for consumers and producers', while the International Energy Agency (IEA) uses the definition that energy subsidies are 'any government action that concerns primarily the energy sector that lowers the cost of energy production, raises the price received by energy producers or lowers the price paid by energy consumers'.

EU legislation requires member states to report regularly to the European Commission on energy policy updates. In fact, the MURE (Mesures d'Utilization Rationelle de l'Energie) is a Commission-funded qualitative database of energy and climate change policy measures and subsidies undertaken by each EU member state. It is this database which shows that the UK has an incredible 63 policy measures aimed at addressing the energy and climate-change agenda.

The Climate Change Levy is one of these measures which taxes the use of energy in industry. The Renewables Obligation (RO) obliges energy suppliers to produce a specified amount of energy from renewable sources and the Feed-in Tariffs provide a fixed payment for electricity generated privately from renewable or low-carbon sources. The RO is the main financial mechanism by which the government incentivises the renewable energy movement. In terms of costs to consumers, the RO alone has risen from £278 million in 2002 to more than £1.4 billion in 2011.

Electricity suppliers meet their obligations by presenting Renewables Obligation Certificates (ROCs) to the regulator, OFGEM. If they do not have sufficient ROCs to cover their obligation, they must make a payment into a buy-out fund that is allocated to the ROCs after deducting OFGEM's costs for administration. In 2009, renewable technologies were 'banded' for ROCs support which provides varying levels of support for different renewable technologies depending on their relative maturity, development cost and associated risk. For 2011/12 the average estimated auction price of a ROC was £46.61

(www.e-roc.co.uk) and suppliers of onshore wind energy were given 1 ROC for every MWh produced. Offshore wind farms receive 2 ROCs per MWh and tidal and wave power installations will be paid 5 ROCs per MWh.

The Feed-in Tariffs (FiTs) are another of the methods which drive the rush for renewables. The Department of Energy and Climate Change introduced the FiTs in 2010 to incentivise small-scale (less than 5 MW) low-carbon electricity generation. FiTs are mainly used to encourage electricity generation by organisations, businesses, communities and individuals which are not normally involved in the electricity market. The premise is that they invest in generating plant and in return they are guaranteed payment for any electricity they export to the grid for up to 25 years, payment being funded by charges made by the electricity suppliers to their customers. For wind turbines, suppliers receive around 31.5 pence per kWh exported back to the grid.

In a recent policy report prepared for the Global Warming Policy Foundation, Professor Gordon Hughes estimated that the total consumer bill for wind subsidies by 2030 will top a shocking £130 billion. Other reports have exposed that a dozen of the biggest UK landowners will collect almost £850 million in subsidies between them. This is an extraordinary amount of funding that will be paid by ordinary consumers through hidden additions to their electricity bills. But this is not surprising when you consider that for an average industrial onshore turbine, the developer can generate electricity that is worth around £150,000 per year. Concurrently, that same turbine can attract annual subsidies upwards of £250,000. Without these subsidies, it is unlikely that anyone would invest in wind power; but with them, wind farms become a very attractive and extremely lucrative option for developers.

As we have seen, there are currently around 4,000 turbines operating in 355 wind farms in the UK. Another 1,200 turbines are under construction and around 2,000 more have secured planning permission. A further 5,000 are in the planning pipeline. Britain will soon be bristling with these giant turbines. There will be few places left in our countryside from which you will not be able to see a massive turbine and its associated pylons and overhead lines. There are now so many planning applications pending for on- and offshore turbines that it is hard to keep track of them all and the UK and Scottish Governments appear to have no central database tracking their construction. It was estimated in August 2012 that there were over 11,000 turbines either built, under construction or in the planning pipeline across the UK, both on- and offshore.

The installed cost of these 4,000 operational turbines amounts to a staggering £7 billion. At full installed capacity they could generate 6,800 MW, but as we know, they run at around 25% efficiency (load factor), and therefore these 4,000 installed turbines can produce only around 1,700 MW on average. Hunterston B, the ageing nuclear generator on the Ayrshire coast, has a maximum capacity of 1,140 MW and can run at 80% efficiency. A replacement nuclear plant could be constructed for around £3.5 billion including all costs for decommissioning and safe storage of waste.

So, for the same investment, we could build two new nuclear power stations which would supply a constant stream of carbon-free electricity, operate for 60 years rather than 20 and run at 80% efficiency rather than the paltry 25% efficiency of the wind turbines. This is an escalating financial scandal which threatens to undermine the whole energy sector. It is an outrage on a similar scale to the banking scandal and could undermine even further the entire Scottish and possibly even the UK economy.

And rest assured, even once all 11,000 of these turbines are up and running, it will not be nearly enough to achieve the UK target of 20% renewable energy by 2020, or the even more ludicrous 100% electricity generation target set by the Scottish Government. To reach these targets we will need a six-fold increase in giant turbines, or 60,000 of them across the UK, many of them in Scotland. And they are getting taller and taller. There are already 150-metre-high turbines being installed in Scotland and there is now talk of bringing in 200-metre giants.

Those who claim that offshore wind developments will provide the answer are also well off track, as a later chapter in this book will explain. The Royal Academy of Engineering has calculated that the cost of a kilowatt hour of electricity produced by an offshore wind turbine is 7.2p, compared to 2.2p from gas, 2.3p from nuclear and 2.5p from coal. The big power companies are no longer farming wind; they are farming subsidies and the poor consumers have to foot the bill.

Additionally, renewable energy developers frequently mislead us about the true cost of their wind farms. This is because the published costs of a wind development do not include associated infrastructure costs which are necessary for a functioning wind farm. The cost of power lines to transmit the electricity, for example, is usually overlooked. In 'Why is wind power so Expensive? An economic analysis', Gordon Hughes reported that it is the National Grid that provides this service and therefore the costs are seen as falling outside the project costs, even though consumers end up paying the bill.

During the coldest days of winter, when temperatures plummet and there is no wind for days on end, Scotland's wind turbines will be standing idle while we shiver in the cold. Between November 2008 and December 2010, wind output mostly from Scottish onshore wind farms had a load factor of just over 24%.

In December 2010, when temperatures fell below -15°C in many parts of the UK, wholesale electricity prices surged. Peak demand for the UK on 20 December 2010 was just over 60,000 MW. Yet, because there was virtually no wind, energy produced by all of our installed wind turbines contributed a pathetic 52 MW. Despite billions of pounds of investment and limitless subsidies, our wind turbine fleet was producing a feeble 2.43% of its own capacity – and little more than 0.2% of the nation's electricity needs.

That is why every single megawatt of generating capacity from our onshore and offshore wind farms will need to be matched by a back-up supply from high CO_2-emitting fossil-fuel gas or coal-fired plants, which nobody in the Scottish Government is planning for. And since the SNP Government has already ruled out any new nuclear plants in Scotland, we can be certain of one thing: we'll have to import nuclear-generated electricity from England, or the lights are going to go out.

The relentless rise in electricity and gas bills by all of the major power companies is directly related to this renewable madness and these big power companies are reaping the benefits at the public's expense. One of the biggest renewable energy companies – Scottish and Southern Energy (SSE) – saw its pre-tax profits rise in 2011 to a massive £1.34 billion. The big power companies are raking it in from the consumers and posting profits that have soared to billions every year.

According to evidence given to the House of Commons Energy and Climate Change Committee in July 2012, by Professor Gordon Hughes of the Global Warming Policy Foundation, 'When wind power is available, its low operating cost and market arrangements mean that it displaces other forms of generation. Market prices are lower, so that other generators require higher prices during periods of low wind availability to cover their operating and capital costs. It is expensive and inefficient to run large nuclear or coal plants to match fluctuations in demand or wind availability, so that their operating and maintenance costs will be higher. At the same time, the risks of investing in new generating capacity will be increased by the impact of wind power on market prices, so that the cost of capital will be higher. Even if wind power was no more expensive per MWh than power from other sources its impact on other generators would still increase the aggregate cost of meeting the UK's electricity demand, probably by a substantial margin.'

Professor Hughes told the House of Commons committee that 'There is no escape from the consequences of the impact of wind power on other parts of the electricity system.' He continued: 'Meeting the UK Government's target for renewable generation in 2020 will require total wind capacity of 36 GW backed up by 21 GW of open-cycle gas plants plus large complementary investments in transmission capacity. Allowing for the shorter life of wind turbines, the investment outlay for this wind scenario will be about £124 billion. The same electricity demand could be met from 21.5 GW of combined cycle gas plants with a capital cost of £13 billion – this is the gas scenario. Wind farms have relatively high operating and maintenance costs but they require no fuel. Overall, the net saving in fuel, operating and maintenance costs for the wind scenario relative to the gas scenario is less than £200 million per year, a very poor return on an additional investment of over £110 billion.'

Professor Hughes informed the committee that: 'There is a significant risk that annual CO_2 emissions could be greater under the wind scenario than the gas scenario.' He continued: 'Wind power is an extraordinarily expensive and inefficient way of reducing CO_2 emissions when compared with the option of investing in efficient and flexible gas combined cycle plants. Of course, this is not the way in which the case is usually presented. Instead, comparisons are made between wind power and old coal- or gas-fired plants. Whatever happens, much of the coal capacity must be scrapped, while older gas plants will operate for fewer hours per year. It is not a matter of old-versus-new capacity. The correct comparison is between alternative ways of meeting the UK's future demand for electricity for both base and peak load, allowing for the back-up necessary to deal with the intermittency of wind power.'

The conclusion of Professor Hughes' evidence to the House of Commons Energy and Climate Change Committee was that: 'Wind generation imposes heavy costs on other parts of the electricity system which are not borne by wind operators. This gives rise to hidden subsidies that must be passed on to electricity consumers. In the interest of both transparency and efficiency, wind operators should be required to bear the costs of transmission, storage and back-up capacity needed to meet electricity demand. Only then will it be possible to get a true picture of the costs and benefits of relying on wind power rather than alternative ways of reducing CO_2 emissions.'

Notwithstanding all of the available evidence, a common cry

by the pro-wind lobby is that all types of energy are subsidised and coal gets more subsidy than wind. In an article in *The Guardian* on 27 February 2012, the headline read 'Wind power still gets lower public subsidies than fossil fuel tax breaks'. The strapline stated: 'Financial support for renewable energy under attack as gas, oil and coal still subsidised to a far greater extent, new data shows.'

Stuart Young Consulting was engaged by CATS – Communities Against Turbines Scotland – to examine the source and validity of these claims. Stuart Young's findings, published in August 2012, make interesting reading. His report begins by stating: 'The claim in the *Guardian* article has no substance or merit. It is disingenuous, totally misleading and is predicated upon a highly inventive and dubious notion of "subsidy".' The report continues: 'The £3.63 billion "fossil-fuel tax breaks" referred to in the article turn out to be an imaginary relief on a level of taxation on domestic fuel which never has and almost certainly never will be levied. On the other hand, the £1,780 million subsidy to wind with which it was compared is indeed a true subsidy to assist an unreliable technology.'

Stuart Young points out that this £1,780 million subsidy to renewables is currently paid for by every man, woman and child in the UK – 66 million people therefore share this burden which works out at £27 per head of population or an average £65 per household in 2012 alone. He further points out that these charges on everyone's power bills are subject to 5% VAT and to the profit margin charged by electricity suppliers.

Robbing the poor to pay the rich

Why are British landowners so keen to have their farms and estates industrialised? Again, it would appear, money is the driver. Rental payments vary and are top secret but it is estimated that a dozen or more of Scotland's wealthiest private landowners will pocket around £1 billion in rental fees over the next eight years.

Based on estimates, the Duke of Roxburghe could net around £1.5 million a year from his 48 120-metre-high turbines at Fallago Rig in the beautiful Lammermuir Hills. Sir Alastair Gordon-Cumming could be earning around £435,000 annually from 29 giant turbines on his Altyre Estate near Forres in Moray. The Earl of Seafield could get £120,000 a year from eight turbines on his estate near Banff. The Earl of Moray is estimated to receive £2 million a year from 49 turbines at Braes O'Doune near Stirling, which are clearly visible from the iconic Stirling Castle. The Earl of Glasgow, a Lib-Dem peer, has 14 turbines on his Kelburn estate, so could be earning upwards of £300,000 a year (the *Herald* 22/08/2011; the *Telegraph* 21/08/2011). And Crown Estates, one of the richest bodies in the country, will net billions of pounds from leasing large tracts of seabed for offshore wind developments.

The potential profits for the fat-cat energy companies are even greater. A single turbine can generate more than £13 million profit over its 20-year lifespan, only half coming from the electricity it produces and the other half from consumer subsidies added directly on to electricity bills.

And the energy companies are even paid for NOT producing electricity. The money just pours in, whether the turbines are operating or not. The latest figures show that the total cost of wind energy constraints for 2011 reached over £25 million (*The Times*, 18 January 2012). That means £25 million of UK and Scottish consumers' money has been paid out to energy companies to keep their turbines switched off because the grid was overloaded! This is quite simply scandalous. Incidents like this only confirm that wind power is unpredictable, intermittent and cannot be trusted. Now OFGEM and the National Grid are predicting that paying wind farms to shut down could cost UK consumers almost £300 million a year by 2020.

So where is all this money coming from? If farmers and landowners are to be paid bloated rentals and energy companies and smaller electricity producers continue to pocket generous ROCs and Feed-in Tariffs, who is footing the bill? Well it should come as no surprise to learn that it is the consumers themselves who have to cough up. In 2011, the Renewable Energy Foundation suggested that the UK's renewables programme would cost the consumer around £6 billion annually. This figure matches those in the Renewable Energy Review, published by the UK Government's Climate Change Committee in May 2011. This review estimated that renewables policies would add 2p/kWh, or £6.5 billion, on to the national electricity bill by 2020.

International experience to date has demonstrated that industrial wind power is simply not viable without heavy subsidies and inflated Feed-in Tariffs. In every country where wind turbines have been installed they have failed to demonstrate economic feasibility; they have failed to demonstrate viability as a solution to global warming; they have failed to achieve significant CO_2 reductions and they have failed to provide efficient electricity production or protection of the environment.

Indeed in countries where industrial wind power has been added to the grid in any volume, consumer costs have rocketed. The two countries with the highest numbers of installed commercial wind turbines, Germany and Denmark, now have the highest electricity bills in Europe. And yet in Germany, *Der Spiegel* reported in a recent article that despite 20,000 installed turbines, CO_2 emissions have not been reduced by even a single gram, because additional coal-burning plants have had to be built to support wind power.

In the UK, the introduction of destabilising wind energy to the grid has meant extensive resort to gas-burning facilities and greatly increased consumption of gas. This has driven up the price of gas dramatically. In Spain, Juan Carlos University laid blame for the country's worsening economic crisis on the wind industry. The report states that the surging price of electricity has driven most of Spain's large energy consumers out of the country.

Meanwhile, the European Investment Bank (EIB) is providing up to €1 billion for the construction of wind turbines and other renewable energy projects in the UK every year. While this is in accordance with the EU's climate-change strategy, it is deeply alarming that the current criteria governing EIB funding lacks transparency and accountability when it comes to examination and due diligence of the projects concerned.

Indeed the EIB is providing €6 billion in total for wind farm developments across the EU, but they simply accept the applications for funding from the member state governments concerned without detailed scrutiny of any of the individual projects. Their view is that EU member states have to achieve climate change targets set by the European Commission and therefore all applications for renewables must be soundly based and can be relied upon to achieve those targets. Sadly, as we know, this is seldom the case, but money continues to be dished out by Brussels to anything that ticks the renewable energy box.

Fuel poverty

What we are witnessing is in fact a dramatic transfer of money from the poor to the rich; from the beleaguered consumers to the wealthy estate owners and power companies. The Department of Energy in Whitehall has revealed that rising bills have pushed 5.5 million UK households into fuel poverty – that's one fifth of British homes.

According to a Scottish Government study, families' fuel bills are rising three times faster than wages. Average fuel bills have reached £1,402 a year, up nearly 62% since 2003. But earnings have only increased by 19% over the same period. The 'Fuel poverty evidence review', published in August 2012, predicted that average fuel spend in Scottish homes would exceed 12% of income by 2015. Families are deemed to be 'fuel poor' if they spend over 10% of net income on household fuel bills. The Scottish Government's fuel poverty report was published, ironically, on the same day as German energy giant E.ON announced that its half-year profits had tripled to £2.45 billion, up 23.7% on its UK operation.

With much colder weather in Scotland, the situation is even worse. Around 900,000 Scottish households are now suffering actual fuel poverty. The most vulnerable people in society are being forced to make the choice between food and fuel. Consumer Focus Scotland has noted that in 2001, the average dual-fuel bill in Scotland was £518. Current estimates suggest that the UK annual dual fuel bill is now £1,258. Now Britons will face an extra £500 on their fuel bills over the next four years to cover the extravagant costs of renewable energy.

The broad definition of fuel poverty is the inability to heat a home to an acceptable standard at a reasonable cost. As mentioned above, the definition used by the Scottish Government is: 'a household is in fuel poverty if, in order to maintain a satisfactory heating regime, it would be required to spend more than 10% of its income on all household fuel use. Extreme fuel poverty is defined as being required to spend 20% of income on fuel to maintain an adequate heating regime.' It is those people on fixed and limited incomes who are most affected by fuel poverty, normally the elderly, students or low-income families.

In Scotland, the eradication of fuel poverty is required by the Housing (Scotland) Act 2001. Although the SNP Government renewed its commitment to eradicating fuel poverty, it actually reduced fuel poverty expenditure from a total of £70.9 million in 2010–2011 to a total of £48 million in 2011–2012. The fuel poverty rate in Scotland did fall from 35.6% in 1996 to 13.4% in 2002. However from that point onwards, the rate has been steadily rising year on year to 32.7% of households in 2009 – so we are almost back to the 1996 levels. These figures for Scotland are significantly higher than the rest of the UK and even dwarf the European average where only 8% of households battled fuel poverty in 2009. Fuel poverty in Scotland just isn't being addressed properly. Is it a coincidence that this rise has coincided with the widespread roll-out of wind power?

In 2007, the SNP put out a policy statement which said: 'We have more than enough energy to end fuel poverty. The SNP will deliver more streamlined government with a greater focus on achieving strategic targets. It will be well-placed to reduce levels of fuel poverty across Scotland and have the breadth of focus and range of responsibilities needed to act effectively.' Now they have pledged to ensure that by November 2016, nobody in Scotland will live in fuel poverty. But achieving a zero fuel-poverty target by 2016 will be impossible while the various market incentives for renewable energy continue to force up costs.

The Scottish Government refuses to face up to the unpalatable truth about renewable energy and therefore they are out of step with other governments in Europe. Italy's government has passed a decree ending subsidies for wind turbines. The centre-right Partido Popular Government in Spain has also announced an end to all subsidies for wind power, stating that it could no longer afford to finance this inefficient method of electricity generation at a time of severe economic austerity. The Dutch Government has also pulled the plug on any further onshore wind developments due to their high costs to consumers and technical problems integrating these sources into the existing infrastructure.

Now the Department of Energy and Climate Change at Whitehall has confirmed that electricity bills are set to double over the next 20 years due to government plans to invest over £110 billion in 'low-carbon power generation'. It is estimated that this investment in a new fleet of nuclear power plants, giant offshore wind turbines and other renewable energy projects will add up to an extra £1,000 a year to the average household electricity bill. Costs of electricity to business and industry will also soar.

Governments across the EU are waking up to the unsustainable costs of wind power as Scotland suffers the consequences of a chaotic and dysfunctional energy policy. At least the coalition government in the UK appears to be beginning to take notice. When Chancellor George Osborne hinted that he might slash government subsidies for wind farms by up to a quarter, there was a panic reaction from the renewables sector and their 'green' supporters. The Lib-Dem coalition partners eventually persuaded Osborne to opt instead for only a 10% cut in subsidies, with a review in 2014. Environment groups say that cuts of 25% will put an end to the development of further wind-power sites, so this is a target which many politicians are now pushing for, including over 100 Tory MPs who wrote to Prime Minister David Cameron in the spring of 2012, urging him to slash subsidies for wind farms. This is now a hot political issue.

The *Sunday Telegraph* (17/06/2012) reported that a senior Conservative source had said subsidies for onshore wind and solar panels are expected to be phased out by the end of the decade. 'This is now very much the direction of travel,' he was quoted as saying. In the same article, Chris Heaton-Harris, the Tory MP who organised over 100 of his Westminster colleagues to sign a letter demanding cuts in subsidy to wind farms, stated: 'I struggle to see how anyone can argue for a policy that gives huge sums of money to big landowners and the big six energy companies, whilst at the same time it thwarts growth and forces tens of thousands into fuel poverty. This policy is not green, progressive or sensible.' He continued: 'The Chancellor should take an axe to these subsidies as soon as possible.'

Bribery

One of the less attractive features of the wind rush in Scotland has been the tendency for the big energy companies to offer 'community benefit' payments to the unwitting citizens who have to live with the visual, environmental and health impact consequences of an industrialised landscape. In any other business environment, such payments, offered always during the planning process and designed specifically to buy off protests, would be considered as bribes. Those offering and accepting such payments would in other circumstances normally be charged with a serious criminal offence. But in the fantasy world of burgeoning wind farms, such payments are commonplace.

To begin the bribery process, developers will prepare a 'questionnaire' to distribute to residents of areas that will be affected by wind farms. The developers hire prominent consultancy and PR firms to prepare the questionnaires and send them to affected communities under the premise that it is designed to find out how each community would like to benefit from a wind farm development. These questionnaires ask respondents to choose between annual payments, one-off payments, private share offers or community share offers. Put simply, they ask people how they would like to be bribed. In a particularly sinister move, the questionnaires also ask respondents what they like about the area and living in the community, as well as their thoughts on wind farms. From past experience, we know that wind farm developers rarely spare a thought for affected residents so this can only be to gauge what level of opposition they will encounter.

Ramblers Scotland recently criticised Scottish and Southern Energy for announcing a £50 million 'community benefit' scheme in the Highlands over the next 25 years, coinciding with their submission of planning applications for two new industrial wind farms. The applications were for 83 turbines at Stronelairg, east of the Glendoe hydro-electric scheme at Fort Augustus, and a further 36 at Bhlaraidh, north-west of Invermoriston, both scaled-down proposals from previous applications that had given rise to fierce local opposition.

The Mountaineering Council of Scotland claimed that the proposals would lead to the virtual industrialisation of large parts of the unspoilt uplands around Loch Ness, while Helen Todd, development officer of Ramblers Scotland, said: 'The level of financial support that developers are able to offer local communities to support wind farm developments is little short of bribery to get planning approval. As we have said before, the integrity of the planning system is at stake' (the *Herald* 30/06/2012).

But these bribes are offered to local communities early in the planning process in a bid to stifle protest. In the case of the Viking wind farm in Shetland, a special Shetland Charitable Trust will be given £20 million every year 'to support community enterprises'. Similarly, the hugely controversial proposal to construct ten giant turbines at Ard Ghaoth near Drymen was backed by a community inducement of £7.2 million, one of the biggest in Scotland, where the developer – Banks Renewables – stated that the money could be used for charities, youth groups and voluntary organisations.

The galling point behind these so-called 'community benefit' payments is that the money is coming directly from the consumers themselves, through costs that are passed directly down the line to the public. The supposed generosity of the energy companies is a charade. Indeed it is a scandal that they can pick and choose who should gain from such bribes in any community, always ensuring that payments are directed only where support for their wind farms can be guaranteed. Such tactics divide communities, pitching those who see the lure of large sums of cash as a literal windfall for their local community council or sports club, as opposed to those who deplore the destruction of their treasured landscape, or the impacts on health and property values.

Although community groups often cannot believe their luck at the apparent vast extent of these so-called community benefits, what they are frequently unaware of is the derisory nature of the sums they are offered in comparison to the huge gains the developers stand to make. Developers and landowners can pocket so much money that they can afford to dangle significant financial carrots under the noses of potentially angry community groups, buying their support and their silence. Little do the community groups realise that these sudden windfalls are coming directly out of their own pockets by way of relentless increases in their electricity bills.

CHAPTER FIVE

Impacts on Landscapes and Tourism

The turbine construction phase

It is the impact of giant industrial wind turbines on the landscape, one of our greatest Scottish assets and one on which our burgeoning tourist industry depends, that gives most cause for concern and has led to an unprecedented increase in public opposition.

We are all familiar with architectural follies like the National Monument on Calton Hill in Edinburgh, or McCaig's Tower in Oban. Scotland has quite a few which have become well-known and prominent landmarks. They were erected by wealthy landowners as symbols of their wealth and power, although they usually have no practical purpose. Now we have a new generation of follies bristling across the length and breadth of our country – industrial wind turbines. Again they have often been erected by wealthy landowners and again they have little or no practical purpose, but this time they have no architectural merit either.

Typically, a foundation for an onshore turbine will be approximately 15–20 metres in diameter and located between 1.2 metres and 2 metres underground. Each giant turbine requires a reinforced concrete foundation around the size of an Olympic swimming pool, utilising around 1,000 tonnes of concrete. The excavated hole is first filled with sand aggregate and cement transported to the site in large lorries along specially constructed access roads.

The exact measurements of the foundations will depend on the turbine selected and the underlying bedrock. Reinforced concrete spread footing is the most common type of foundation, as it is the most simple and economic. It is commonly used in areas with good soil bearing capacity. If the turbine is to be constructed where the soil is weak or the site has uncommon conditions such as constraint of space, or if the terrain is sloping or involves offshore construction, other types of foundation are used. These include pile foundations, drilled shafts and caissons.

Turbine towers are mostly tubular and made of steel, generally painted light grey. The blades are made from fibreglass reinforced polyester or wood epoxy. Taller towers are built up in sections determined by transport and lifting constraints. For the larger towers, which have a diameter at the base big enough to accommodate a double-decker bus, specialist transporters are required to carry the sections from the fabrication yard to the construction site. Typically, these turbines have lifts inside which can take an engineer right up to the nacelle, where necessary maintenance and repair work can be undertaken.

The amount of concrete and steel needed for the construction of every single turbine is enormous. It is worth remembering that the concrete industry is known to be the biggest man-made source of CO_2 on the planet, accounting for around 7% of the world's total emissions, so the carbon footprint of every wind turbine is colossal! In fact each 1,000-tonne turbine plinth will be responsible for approximately 900kg of CO_2 emissions into the atmosphere.

And that is only emissions from the concrete. According to the World Steel Association, steel represents an average of 80% of all materials used to construct a wind turbine. Steel production is also a major source of CO_2. And even after the steel towers have been fabricated, a fibreglass nacelle is installed to house the main drive shaft, gearbox and blade pitch and yaw controls. A typical nacelle for a large turbine will weigh around 10,000kg. A typical fibreglass blade meanwhile will be around 45 metres in length and weigh approximately 1,133kg.

Before any turbines can be constructed at the site of a proposed wind farm, access roads have to be built. These roads must be wide enough and strong enough to carry the huge transporters which haul the component parts of the turbines to the construction site, as well as cranes and heavily laden concrete and aggregate trucks. The aggregate needed to build these roads is usually dug out from the sides of surrounding hills. These are misleadingly called 'borrow pits', suggesting that whatever is taken will be returned, although this is never the case.

When it comes to decommissioning a turbine, usually around 20 years after its erection, it is normally up to the local authority who gave initial planning approval to have a clause in that consent covering decommissioning. These clauses typically require all visible traces of the wind farm to be removed. The concrete foundations should also be removed, but developers often argue that it causes less environmental damage to leave them intact and undisturbed. Bonds are sometimes deposited by the developers at the time of applying for planning consent, for the purpose of meeting the costs of decommissioning.

But more often the reality is that by the time a wind farm has reached the end of its productive life, the renewable energy company responsible for its original construction will have disappeared and local councils will be forced to pursue landowners for the restoration of the wind farm site. This has certainly been the case in the United States. In addition, access roads built over peat bogs do permanent damage to peatland, which can never be restored or remedied; borrow pits are left as gaping quarries; concrete foundations are covered with a thin layer of peat or soil and could remain in place for millennia. When asked about decommissioning at a public meeting in Ayrshire, a spokesman for ScottishPower stated that the wind farms would be handed over to the local community at the end of their productive lifespan. This announcement was met with hoots of derision from the assembled audience.

Landscape vandalism

Cameron McNeish, one of the UK's most renowned mountaineers, has attacked Scottish Government policy on wind energy, stating that the spread of giant turbines and pylons needed to carry their power to the National Grid was 'scarring Scotland'. (*Sunday Times* 08/04/2012). McNeish, who is vice president of Ramblers Scotland, said that the growth of wind farms was causing environmental degradation that was threatening the future of Scotland's tourist industry. He said: 'They will spoil people's enjoyment of the hills for recreation and tourism will suffer.'

McNeish said that in his opinion the Scottish Borders had 'reached saturation point', as had the countryside between Inverness and Speyside, due to the spread of industrial wind turbines. He was voicing his fears in the immediate wake of a warning from three of Scotland's leading conservation and environmental organisations – the John Muir Trust, Ramblers Scotland and the Mountaineering Council of Scotland – who warned of the severe impact on Scotland's landscape and rural communities due to the proliferation of wind turbines.

Their combined concern was highlighted by a press article (*Daily Mail* 24/03/2012) pointing out that over one third (34.7%) of total land in East Renfrewshire had been earmarked by the council for wind farm development. East Renfrewshire Council was closely followed by Highlands at 30% and Scottish Borders at 13%. The local government body Heads of Planning Scotland (HoPS) warned of a 'wind farm landscape' across the country. In fact in June 2012, the Mountaineering Council of Scotland called for a moratorium on commercial wind farms due to their threatening encroachment on Scotland's Munros (mountains over 3,000ft) and Corbetts (between 2,500ft and 3,000ft). Ron Payne, the Mountaineering Council of Scotland's director of landscape and access, was quoted: 'The mountains and wild places of Scotland are a national asset beyond price, yet they risk being irrevocably damaged by commercial wind farm developments' (MCofS 15/06/2012).

A Panelbase poll for the *Sunday Times* and Real Radio Scotland (*Sunday Times* 29/07/2012) found that 57% of voters believe onshore wind farms should never be permitted in sensitive wilderness areas. Only 9% of people polled said they should be allowed. In addition, 37% of Scots said that there should be buffer zones around communities to protect them from turbine projects.

But despite these poll findings, it seems that nowhere in our pristine landscape is safe from the avaricious attention of the industrial wind developers. From Shetland to Lewis; from Dava Moor on the edge of the Cairngorms National Park, to the shores of Loch Ness; from the beautiful Scottish Borders to the unspoiled Ayrshire coast; from Machrihanish to Tiree, nothing is sacred. Scotland has around half of the UK's installed wind turbines on its territory right now. But we also have more than half of the UK's most beautiful landscapes.

The government has given approval to the controversial Beauly–Denny overhead lines, mentioned in an earlier chapter of this book, based on the necessity to upgrade the infrastructure to carry wind-generated energy to the National Grid. The giant pylons involved in this upgrade, each the height of a 15-storey tower block, will pass within a field's length of the iconic Wallace Monument near Stirling, built to commemorate William Wallace ('Braveheart'), who fought for Scotland's independence. They will be clearly visible from Stirling Castle. It is a great irony that this icon of Scottish nationhood is now being besmirched by a nationalist government.

But these are not the only historic sites that will fall foul of the turbine tumult. In 2011, plans were unveiled for a monstrous wind farm development overlooking Urquhart Castle on the shores of Loch Ness. A series of planning applications would have led to more than 150 giant turbines sprouting across the moors around Loch Ness, Scotland's most famous loch, renowned worldwide and a major source of tourist revenue. The local community was up in arms and after determined protests, Scottish and Southern Energy (SSE) revised their plan to include 83 giant turbines on one side of the loch near Fort Augustus, while a further 36 turbines will be erected on the other side of the loch. SSE claims that neither scheme will be visible from the loch, but fierce opposition to the project continues.

In late 2011 I addressed an anti-wind-farm rally in Inverness, which was packed with mountaineers, hill-walkers, cyclists, ramblers, hoteliers, B&B operators, all of whom were appalled at the visual pollution of our landscape which they say is wrecking Scotland's tourist trade. They were echoing similar protests being heard around the country.

Nowhere is safe. In Dumfries and Galloway, ScottishPower Renewables are planning to build up to 19 wind turbines at Ae Forest. Besides the devastating impacts on the forest and wildlife, the wind farm proposal has already dealt a death blow to the world-renowned Tharpaland International Retreat Centre, which attracts tourists from all over the globe. The retreat centre has been functioning successfully for 27 years, but the monks there have decided that wind farms are incompatible with the Buddhist way of life and that Tharpaland would be unable to continue functioning as a retreat and will therefore have to close completely. They have decided to move out of the area and to resettle near Berlin in Germany, claiming that nowhere in Scotland is safe from the wind farm onslaught. The Tharpaland monks are the first of a wave of wind-power refugees in Scotland.

There are even plans to dwarf the famous A-listed Bell Rock Lighthouse in the North Sea. The Lighthouse was built between 1807 and 1810 by Robert Stevenson (Robert Louis Stevenson's father) on the Bell Rock, 11 miles off the Angus coast. It stands 35 metres high and has been the subject of paintings and postcards for the past two centuries. But now giant 60-metre-high wind turbines, some located only five miles from the lighthouse, will tower above it, dominating the seascape. Seagreen Wind Energy is developing the largest of the wind farms in the Firth of Forth Zone, with up to 150 giant turbines located 17 miles from the Bell Rock. Two smaller projects – Inch Cape from Spanish-owned Repsol Nuevas Energias and Neart na Gaoithe from Mainstream Renewables – will be positioned between five and seven miles from the lighthouse respectively.

A 2008 study entitled 'The Economic Impacts of Wind Farms on Scottish Tourism', conducted by Glasgow Caledonian University for the Scottish Government, showed that the development of onshore wind farms will damage the tourist sector in Scotland. Of tourists questioned, 68% said they wanted a wind-farm-free landscape. The survey even found that tourists would be prepared to pay up to 25% extra to have a view unspoiled by wind farms from their hotel or B&B accommodation.

Impacts on the film industry

The report proved that tourism and wind farms are just not compatible. Indeed, we can forget our some of our fabulous Scottish landscape featuring again in blockbuster movies. There is no place for industrial wind turbines and pylons as a backdrop to a Jacobean melodrama! Nor can you have Rob Roy taking shelter from the redcoats behind a giant steel windmill! Scotland's USP is its scenery. Destroy it and you destroy the tourist trade.

Rhoderick Noble, Factor of the Ardverikie Estate on the shores of Loch Laggan, east of Fort William, said: 'The movie *Kidnapped* went to New Zealand for the very reason that the producers felt it was no longer possible to film in the Highlands because of the proliferation of power lines, and Douglas Rae, producer of *Monarch of the Glen*, which was based here at Ardverikie, has stated publicly that the series would not have been set here if the proposed pylons (Beauly–Denny) were in place.'

Mr Noble continued: 'There is no doubt that *Monarch of the Glen* gave a tremendous boost to the area and brought about an increase in tourism. There were guest houses in Newtonmore that were full from March to October for seven years with the cast and production crew from Monarch' (*Strathspey & Badenoch Herald* 25/10/2006).

In March 2012, there was widespread media coverage that Oscar winners Russell Crowe and director Ridley Scott were considering filming the sequel to their blockbuster hit *Robin Hood* at the Clanranald Trust site in Carron Valley, Denny, Stirlingshire. However, the Trust's manager highlighted that plans to build 16 120-metre-tall turbines at the nearby Cairnoch Hill site would damage the unspoilt countryside and undermine the film project.

Don't VisitScotland

With all of this going on, it is astonishing to read the words of Malcolm Roughead, Chief Executive of VisitScotland. Welcoming in April 2012 the results of a poll of 2,000 UK residents, which claimed the presence of a wind farm would have little impact on a tourist's decision to holiday in Scotland, he stated: 'We sell Scotland to the world, bringing millions of visitors to the country and boosting the economy by billions of pounds. The visitor experience is therefore a huge priority for us – we know visitors come here for the scenery and landscapes and our marketing activity works hard to promote those aspects. And so we are both reassured and encouraged by the findings of this survey which suggest that, at the current time, the overwhelming majority of consumers do not feel wind farms spoil the look of the countryside.' This staggering complacency from a person whose job it is to promote Scotland as a tourist destination must surely shock people in the industry.

VisitScotland should take care. Its attempts to bend over backwards to comply with the wishes of its main paymaster – the Scottish Government – may not entirely reflect the views of Scotland's leading tourist venues. A spokesman for the luxury Gleneagles Hotel in Perthshire was quoted in the *Telegraph* (30/06/2012), stating: 'Guests expect to see an unspoiled landscape, as this is what they associate with Scotland.' Instead they notice a 'marked visual impact' from nearby wind farms. 'Cumulative and rapid growth of more wind farms threatens the very essence of why these places are so well-regarded and known.'

Indeed, current government policy flies in the face of previous advice. Glasgow Caledonian University's research for the Scottish Government stated that: 'Numerous surveys have established the importance of the Scottish landscape to potential tourists to

Scotland. It has also long been realised that many people find that man-made structures such as pylons and wind turbines reduced the attractiveness of a landscape. A reduction in quality of an important feature must inevitably reduce demand which will result in either reduced prices or reduced numbers or both. This loss of expenditure will lead to a reduction in economic activity and result in a loss of income and jobs. The question therefore is not whether wind farms have an economic impact but rather what is the likely size of that impact, a far more difficult question to answer.'

Perhaps VisitScotland's Chief Executive should ask the pertinent question – what impacts will giant industrial wind turbines towering above Loch Ness, Stirling Castle, the Wallace Monument, Culzean Castle, the Old Course at St Andrews and the Turnberry Championship courses have on tourist income? If he seriously expects anyone to believe that the answer is 'none', then perhaps it is time VisitScotland sought a new boss or perhaps changed its name to 'Don't VisitScotland'!

However, maybe the worm is beginning to turn at last. In October 2012, VisitScotland finally broke cover and lodged an objection to a proposed wind farm near Lockerbie, Dumfries and Galloway, complaining that it could have a detrimental impact on walkers on the Southern Upland Way.

Indeed, Professor Jane Bower, Vice Convener for the Association for the Protection of Rural Scotland, has taken a far more realistic approach, stating that Scotland should rethink its 'headlong dash for antiquated and inefficient wind turbines'. Her antipathy to wind has been echoed south of the border by one of her counterparts, the Chairman of the Council for the Protection of Rural England (CPRE), who said that a 'dramatic proliferation' of wind turbines is blighting the English countryside.

In a May 2012 interview with the *Telegraph*, Sir Simon Jenkins, Chairman of the National Trust, warned that wind was the 'least efficient' form of green power, and risked blighting the British landscape. 'We are basically against these things wherever they might intrude upon the landscape,' he said.

Why is it that the English are stepping up to the plate to protect their countryside and landscape heritage, while Scots in similar positions of authority are remaining silent or providing lickspittle, fawning support for a policy that is trashing our unique selling points? It seems that the John Muir Trust is amongst a minority of organisations that care about conserving Scotland's wild land. They have pointed to a recent report by Scottish Natural Heritage ('Natural Heritage Indicators: Visual influence of built development and land use change'), which shows how the total amount of land in Scotland unaffected by any form of built development has shrunk dramatically from 41% in 2002 to only 31% by the end of 2008. It then declined by a further 3% in 2009.

SNH stated to the Scottish Parliament Public Petitions Committee that: 'Our initial analysis suggests that the most significant contributor to this decline is the development of wind farms, a consequence of their prominence and extensive visibility and siting in rural locations with little or no previous development.' The John Muir Trust concluded that: 'At the moment, the rapid expansion of wind farms represents the biggest threat to our remaining wild land.'

As part of an epic poem written in objection to the proposed wind farm at Clashindarroch, Huntly, local poet Donald S. Wright wrote the following quatrain:

> So, scarce a pick o' CO_2,
> Their gran' design will save,
> But smairtly wreck the tourists' view,
> An' gaur the wildlife leave.

The impact on golf

Scotland is world renowned as 'the home of golf' and people travel from all around the globe to play our courses. Scottish Enterprise estimated that golf tourism's contribution to our economy is around £220 million annually, with significant spillover benefits to local economies. So it should come as no surprise that Donald Trump almost pulled the plug on the five-star hotel and associated luxury homes aspect of his multi-million-pound golf development in Aberdeenshire, blaming an offshore wind farm that will be built directly in the line of sight of these buildings. The golf course, which many experts have described as one of the best in the world, has now been built, and in October 2012 Trump announced that he would go ahead with the hotel development after all, but would fight the adjacent windfarm proposal through the courts.

But Donald Trump is right. He has refused to bow to the turbine tyrants in the Scottish Government. He chose the Menie Estate in Aberdeenshire for development because of the fantastic seascape and he was appalled when plans were announced to vandalise this view with the construction of 11 65-storey-high industrial wind turbines only one and a half miles from his luxury hotel.

Golf tourism aids economic growth as private developers like Trump invest heavily in commercial golf facilities. A report into golf tourism by SQW Consulting showed that, in recent years, quite apart from Donald Trump's development, investors have pumped a further £250 million into various developments at St Andrews, Carrick Golf and Spa at Loch Lomond, Machrihanish Dunes, Castle Stuart, Turnberry and Rowallan Castle. This has not only helped generate employment and economic growth in Scotland but has enhanced our reputation for having the finest courses in the world.

The importance of golf to Scotland's economy simply cannot be overstated. We have nine Open Championship venues and 550 golf courses spread across the country. KPMG's Golf Benchmark Survey showed that Scotland has one of the highest levels of supply of golf courses per head of population (one course for every 9,300 people) in the world. We also have one of the highest golf participation rates in the world with 5% of the population registered as golfers in 2007. But it is not the number of courses or golfers which attracts tourists from both home and abroad; it is their high level of quality and spectacular setting, which adds to the unique nature and challenge of playing the country's links courses.

Yet wind turbines are already beginning to intrude on many of these world-class courses. We have seen public outrage at the Kenly wind farm situated near the Old Course at St Andrews. Members of the Royal Aberdeen Golf Club were stunned when a 65-metre turbine appeared near the 14th tee, the flicker effect from which makes playing golf almost impossible (see film on www.youtube.com). Even so, Aberdeen City Council is now considering giving planning approval to a second giant turbine at the same location.

With so many projects under construction, consented and planned, with larger, more obtrusive turbines rising to 120 metres, we should not be asking a handful of people if, hypothetically, they would avoid an area where a wind farm might be built. We should be asking the nation how it would feel when our most valuable resource is actually covered with turbines, and we should be asking the more than 200,000 people employed by the tourist industry how they will cope when the sector collapses under the sheer weight of steel and fibreglass.

It is no coincidence that the publication of new research, claiming that wind farms do not impact on tourism, coincided with Donald Trump's appearance at the Scottish Parliament's Renewables Inquiry in early 2012. But the very fact that Trump, one of the world's most successful hoteliers, flew in especially to give evidence immediately illustrated that the tourist industry, and those involved in it, have already been impacted.

VisitScotland's 'Wind Farm Consumer Research' (2011) surveyed 1,000 Scots and sought their opinions on the importance of natural landscapes and the impacts of wind farms on their holiday choices. Fergus Ewing, the Minister for Energy, Enterprise and Tourism, even commented that the figures prove 'the vast majority of visitors to Scotland do not see wind farms as a problem'.

But what Mr Ewing should have said is that the vast majority of visitors to Scotland do not see wind farms as a problem *yet*. The impacts of our government's mad dash for wind power will not be immediate, despite the rapid spread of turbines across our nation. People's attitudes will undoubtedly change when the dozens of wind farms currently under construction, on top of the ones already in place, are completed. Attitudes will change when the 112 consented projects are begun and by the time work commences on the massive 169 planned wind farms in Scotland, people will realise that our invaluable countryside has become irreparably damaged. But by that stage it will be too late. The damage will have been done.

Along with steep rises in consumer bills, the tourist industry will bear the brunt of the economic turmoil caused by the spread of wind power. Scotland's spectacular scenery is the cornerstone of our tourist industry, which encompasses around 20,000 businesses employing over 200,000 people (8–9% of the workforce) and generating upwards of £4.2 billion annually for our economy. It is why many of the 2.5 million international visitors flock here each year and why 15 million tourists took overnight trips to or within Scotland in 2008. Scottish Natural Heritage estimated that nature-based tourism alone provides 39,000 jobs and £1.4 billion annually. Scotland without our varied natural heritage is unimaginable and building wind farms – or, as Alex Salmond says, 'industrialising Scotland' – will inevitably impact massively on tourism.

Even VisitScotland, in their Tourism Prospectus (2007) noted that: 'Visitors do not primarily come to Scotland because they like hotel bedrooms! They do not come in order to drive up and down the roads, or use the railways and airports. Visitors to Scotland come for an experience that is rooted in our hills and glens, our castles and towns, our history, our culture, our way of life and our people. Visitors participate in any number of activities, pursue many different interests, see many different places, but they do so against a distinctive backdrop that is the country of Scotland.'

That is why it is all the more surprising that the Scottish Government has repeatedly ignored calls for a national strategy for wind farm developments. It would be simple for the government to design a map of Scotland, the way they did in Germany, which showed where wind farms could be built and where they could never be built. But all pleas to have such a coherent policy have been consistently ignored.

So now we have a free for all, where opportunist energy companies, keen to cash in on the lucrative subsidies, submit planning applications for projects in locations that are wholly inappropriate, such as overlooking Loch Ness or Culzean Castle, forcing local councils, communities and concerned members of the public to expend enormous amounts of time and money fighting against plans that would ravage our landscape and destroy our tourist trade.

Such a case arose in South Ayrshire where an English energy company applied for planning approval for a wind farm on the highest hill overlooking Culzean Castle, one of Scotland's most iconic tourist attractions. Culzean was designed by Robert Adam to fit into the surrounding hills and bay as an architectural masterpiece of global importance. Even to consider erecting industrial turbines in this unique location was criminal. But the local council and community had to withstand several years of cost and anxiety as they fought against the plans, until finally the application was withdrawn and the company walked away. Undeterred, a new application is pending for a series of individual turbines dotted across the same hills, so the whole tortuous war of attrition will start again.

In the meantime, South Ayrshire Council had racked up tens of thousands of pounds in costs dealing with this egregious planning application – costs that the local council-tax payers will have to bear. Why are these energy companies not forced to pay compensation in such cases? Why are they allowed to apply to build wind farms in such locations in the first place? The whole system is a shambles and it is Scotland's renowned tourist sector that will suffer.

Somehow, we now have a situation where people feel ashamed to protest against wind farms on the basis that they vandalise our bucolic landscape. The Scottish Government's propaganda machine makes us feel that this is not a valid reason to object. Their climate-change scaremongering has wrongly convinced us that we are doomed unless we give way to hundreds of turbines. Thankfully, more and more people seem to be realising that scarring our globally renowned landscape might have been a bad idea. It is reminiscent of the line in the famous Joni Mitchell song 'Big Yellow Taxi' – 'You don't know what you've got 'til it's gone'.

Nevertheless, public anti-wind-farm sentiment is growing, but people shouldn't be made to feel bad because they want Scotland's natural beauty to remain naturally beautiful. This doesn't mean that they are climate-change sceptics and it doesn't mean that they are against green or renewable technology. It simply means that they want things to be done properly. They know what they have and they don't want to wait until it's gone.

The beauty of Scotland's landscape cannot be described using facts and figures. It has been celebrated in poems, songs and paintings and it is the centrepiece of our tourist industry. Most importantly, it constitutes a major part of our national identity. But this doesn't mean that protecting it should not be a high priority. Why should ancient Scottish scenery be plundered to serve the short-term needs of the government? No single person owns Scotland's beauty. It belongs to us all and we all must protect it. But whilst we have burgeoning public opposition on one hand, we have unprecedented vandalism on the other, with hundreds of new wind farm schemes at either the planning application or submission stage.

We don't need scientific research to tell us that people visit Scotland for our rural landscapes, natural countryside and varied wildlife. Everyone is aware that thousands of tourists flock to Scotland annually to enjoy our great outdoors and our world-class golf courses. Strangely, with hundreds of new wind farms currently under construction, consented or in the planning process, the Scottish Government has billed 2013 as the Year of Natural Scotland, to promote our 'outstanding natural beauty and cultural heritage'. By that time, how much natural beauty will actually be left?

Tourists will only come for so long until they see that what they came for is no longer there. The mere prospect of turbines is already affecting our tourist industry, as we have seen with Donald Trump's refusal to proceed with his five-star luxury hotel and golf development at Menie in Aberdeenshire. When Trump was asked by the SNP's Chic Brodie MSP during the Holyrood inquiry if he had evidence to back up his claims that wind turbines impacted negatively on the tourist industry, the tycoon thumped his chest and roared, 'I am the evidence!'

Trump has now employed a team of expensive lawyers to fight the offshore wind turbine project. But despite his determined campaign to stop the turbines and complete his golf development at Menie Estate, leading SNP politicians have spread malicious rumours that the economic situation had created financial problems for Trump's project and he was secretly looking for an excuse to bail out, blaming the turbines in the process. This attempt to smear Trump has exposed a sinister side to the SNP's strategy of 'it's my way or the highway!'

The Scottish Government's overly ambitious policy is target-driven, it has no apparent ceiling and it has failed to take account of public concerns for Scotland's landscape. They forced it upon us, assuming that we will just get used to the wind farms. Public consultations have been inadequate at best. Developers faced with effective public opposition are increasingly offering 'doucers' to buy off affected communities.

Fergus Ewing, wearing two conflicting hats as minister for tourism and minister for energy, put out an official New Year message in December 2011 in which he boasted that he wants to 're-industrialise our country' so Scotland can become 'a global green energy powerhouse'. These statements are typically dishonest and misleading. How on earth can it be a process of 're-industrialisation' when it targets and destroys pristine natural areas and rural communities that were never industrialised in the first place? This is straightforward plunder of Scotland's unique and diverse natural environment. It is simply madness driven by an obsession for renewables.

Just look at Whitelee Wind Farm, south of Glasgow. It currently encompasses 140 turbines covering 20 square miles with 44 miles of service roads. With three approved extensions, Whitelee will soon be home to 215 turbines. How can such a development even be called a 'farm'? It is, in fact, a huge power station. But how would the public feel if the government decided to build a 20-square-mile coal, oil, gas or nuclear power station? There would be total outrage and demonstrations on the streets. The Greens and environmentalists would be up in arms and yet they are the self-same people who support this massive industrialisation of our Scottish landscape on a scale that has never before been realised.

While I am convinced that the current energy policy was never envisaged in a spirit of maliciousness, nevertheless attempts to force ill-conceived policies like this on an unwitting public are a grave mistake. But even when politicians come to realise the severity of their mistake, it is unlikely that they will admit they were wrong. Politicians seldom do! They fear that a change in policy will lead to a dramatic loss of face and they are facing too much pressure from landowners and energy companies who have become captivated by the huge subsidies fed to them. This is why they are scrambling for ways to continue the mad dash for renewables and now they are casting their eyes offshore with a view to defiling our world-renowned seascapes.

CHAPTER SIX

Impacts on Human and Animal Health and National Security

Impact on birds and animals

The construction of industrial wind farms has a direct impact on wildlife habitats, protected species of birds and ecologically fragile water courses. With the global race on to build mega wind turbines with a 275-metre wingspan, the impact on birds will become particularly significant.

Turbines chop up rare birds, including white-tailed eagles in Norway and golden eagles in California. They will do the same here. Their constant low-frequency noise and vibrations are intolerable to livestock and wildlife. There are innumerable reports of the long-term, irreversible and destructive impact that industrial turbines have on wildlife, leading to abandoned habitats, as well as the negative impact on livestock performance and production.

The massive 5,000 turbine wind farm at Altamont Pass in California has become one of the most controversial amongst conservationists because it was built directly on a major migratory path for birds. The National Audubon Society, which is America's equivalent of the RSPB, estimates that the Altamont wind farm chops up around 10,000 birds every year, including 80 protected golden eagles, 380 owls, 300 red-tailed hawks and 330 falcons. Thousands of bats were also being minced annually by the flailing blades of the massed turbines. Now Altamont has been forced to close down for four months every year during the main migration season, to reduce the annual carnage.

But even elsewhere in America the signs of horrific damage to wildlife are plain to see. On the edge of the Mojave Desert, the Tehachapi Pass wind farm has been blamed for causing the death of hundreds of golden eagles, whose carcasses can be found littering the desert under the big, industrial turbines. It is a great tragedy that these majestic birds, because of their size, find it almost impossible to manoeuvre through forests of enormous turbine blades whirling at speeds of up to 200mph, especially when they are concentrating on searching for prey. Such is the annual carnage that the American Bird Conservancy now estimates that wind farms are killing between 75,000 and 275,000 birds every year.

A similar fate awaits the golden eagles of Angus. Plans to construct 17 134-metre-high turbines at Nathro Hill near Glen Lethnot in North Angus, pose a serious threat to Scotland's most iconic and rare protected birds. The wind farm will be erected on the top of a hill at around 495 metres above sea level. Young golden eagles have been tagged and tracked by satellite all around the proposed wind farm site at Nathro Hill. This is a natural flight boundary between the highlands and lowlands of Scotland for many of the eagles which range for thousands of kilometres around the country each year. The proposed wind farm would be constructed within 500 metres of a 'Golden Eagle Special Protection Area', an area of international importance for the species. The wind farm will, in effect, become a rotary mincing machine for birds across an area of 14 hectares. Eagles flying across Nathro Hill, their eyes scanning the ground for prey in the frequently heavy mist, will have little chance to avoid the massive slicing rotor blades that chop through the air every second.

The proposal for a giant wind array in the Firth of Forth, off the coast at Fife Ness, will pose an imminent threat to the abundant wildlife for which the area is renowned. It is estimated that around 125 turbines will be erected offshore at this location, each probably 200 metres high, with rotor diameters of over 100 metres. By comparison, the London Eye is only 135 metres high! The potential for untold damage to the internationally recognised colonies of seabirds and marine mammals around the Isle of May and the Bass

Rock from the construction and operation of this wind farm is enormous. But this £1.4 billion project will rake in 200% subsidies for the developers, so it is perhaps not surprising that the fate of seals, puffins and dolphins are of such little importance.

Low-frequency noise from wind turbines is a major source of impact on animals too. Animals are even more susceptible to low-frequency noise than humans. They rely on a whole range of sounds that are not audible to humans. In an almost silent rural environment, the intrusive nature of low-frequency noise and vibration into the soil by industrial wind turbines can cause great disturbance in such sensitive habitats, certainly threatening and confusing wildlife. The hearing and vibration sensitivity of most birds and animals is much more acute than human sound and vibration perception.

Small wind turbines, which are increasingly proliferating across rural Scotland, have been found to cut bat activity in their immediate vicinity by over 50%. Academics from Stirling University carried out research funded by the Leverhulme Trust, examining 20 separate sites where small wind turbines had been erected across Scotland. They discovered that bat activity was 54% lower in close proximity to operating turbines compared to those which were stopped (*Herald* 31/07/2012).

Sound confusion can also cause a failure of hunting success, a failure to self-defend and ultimately can raise questions of survival. Similarly, many animals use shadow flicker as a sign of approaching danger. The constant flicker effect from industrial turbines can lead to significant confusion and to habitat abandonment. It can be concluded that wind turbine developments near any important wildlife sites will have a long-term, irreversible destructive impact upon these habitats. The effect will be cumulative and will increase the longer the turbines remain in place.

Impacts on humans

And of course, if animals suffer these dire consequences, it is easy to understand why humans are driven to despair when giant turbines are built near their homes. Low-frequency sounds that travel easily and vary according to the wind constitute a permanent risk to people exposed to them. The effects of such broad-spectrum low-intensity noise, especially at night, combined with shadow flicker and vibration – which can affect individuals indoors as well as outdoors – have caused people to abandon their homes in distress and ill health. As described in Chapter 2 of this book, with the specific cases of Kay Siddell and Norrie Gibson, the profit being made by the landowners and power companies is often at the expense of people's quality of life.

The impact of wind farms on human health has become a hot topic recently in the media and public outrage is growing. Increasing numbers of people report the unbearable acoustic and optical effects of turbines. After long ignoring the problem, governments and renewable energy companies have employed acoustic experts to argue that audible and low-frequency noise from turbines is unlikely to affect health. But independent biomedical experts have come to very different conclusions, showing that living close to a turbine can cause headaches, dizziness, sleep deprivation, unsteadiness, nausea, exhaustion, mood swings and the inability to concentrate.

Government experts argue that turbines don't create any more sound than already exists in rural areas from normal background noise. However, the human ear does not just respond to 'loudness' or sound pressure that is measured in decibels. It also responds to sound frequency which is measured in Hertz and affects the body even when a sound is 'inaudible'. This is low-frequency noise and is a very important factor to consider. Turbine blades emit low-frequency noise which travels easily and varies according to the wind. This constitutes a permanent risk to people exposed to it.

There is even military weaponry that relies on low-frequency sound for crowd-control purposes. At high intensities it creates discrepancies in the brain, producing disorientation in the body and resulting in what is called 'simulated sickness'. The Israeli army uses this technology to cause instability, nausea and headaches. Suddenly, nobody feels like protesting any more! It is great for crowd control as it has no adverse effects – unless you are exposed to it for hours, as you would be if you lived beside a turbine.

Turbine noise presents a further danger when combined with visual effects such as shadow flicker. This compounds the adverse impact on residents and can induce both physical and psychological symptoms. Visual flicker and 'strobing' effects in sunlight, like a pulsing disco, occur at certain times of the day, similar to when you drive past a row of trees with the sun behind them. Night-time flicker can also occur with the rising and setting of the moon. On elevated ridges, tall turbines can cast shadows for thousands of feet, well above any vegetative screening and nearby residents will be exposed to numerous shadow flickers simultaneously. That is, all three blades of each turbine will cause flicker, and the flicker from each turbine will not be synchronised.

Shadow flicker is a relatively new phenomenon that has not before been experienced on such an industrial scale. Therefore few governments have properly legislated for it. In Sweden, they have recognised shadow flicker and introduced a rule that the calculation of shadow flicker should be made for the entire plot, instead of only the window of a building. Needless to say, the Scottish Government has yet to do the same. The health impacts of turbines need to be scrutinised more rigorously. The Scottish Government's failure to do so amounts to a betrayal of trust.

While the impact of wind turbines on human health has been largely ignored in Scotland, this has certainly not been the case elsewhere. In Canada, a leading physician warned the regional government in Ontario of the dangers in 2009. Dr Robert McMurtry, former Dean of Medicine at the University of Ontario and adviser to the Minister of Health, stated in a report to the Ontario Legislature entitled 'Wind Turbine Syndrome' that while there was a lack of epidemiological evidence to establish either the safety or harmfulness of turbines, there had been, nevertheless, 'many reports of adverse health effects'. His report noted that by September 2009 there had been an 85% increase in the number of

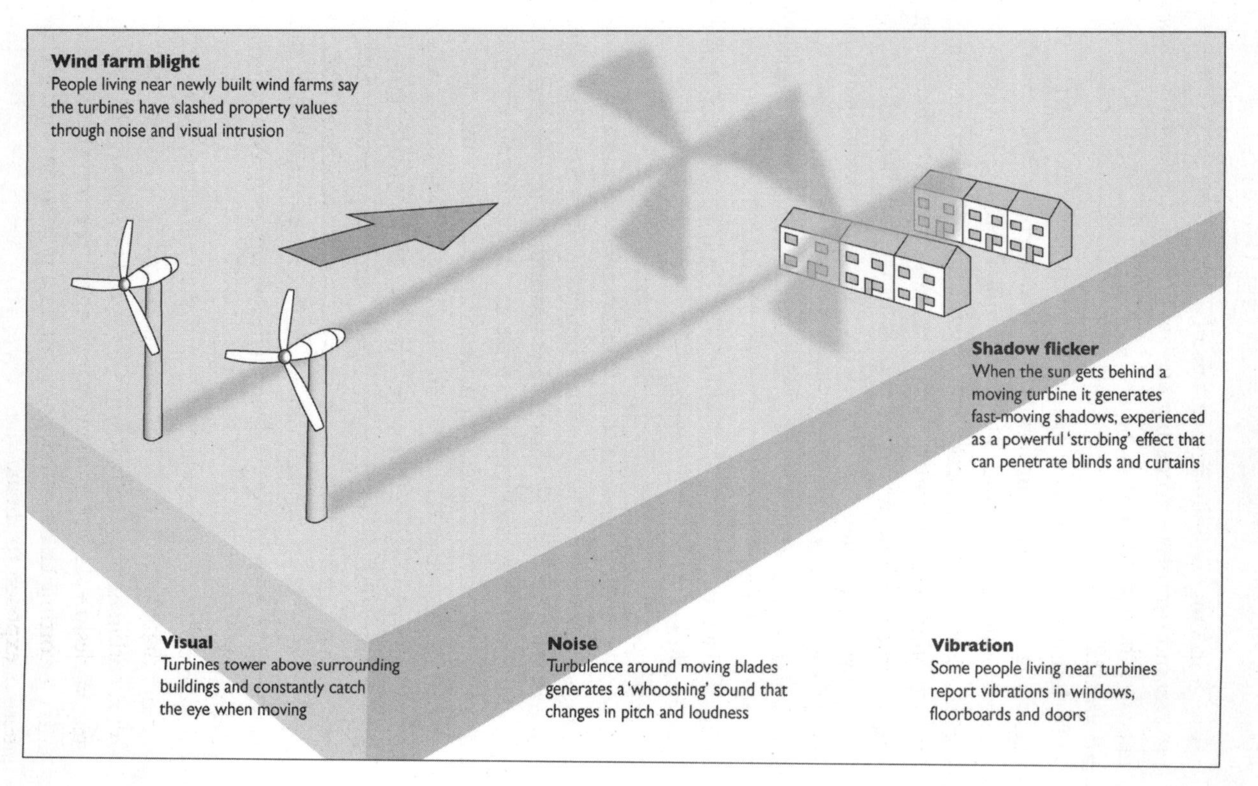

Wind-farm blight

people in Ontario reporting adverse health effects due to industrial wind turbines. 'Some families have been driven from their homes,' his report stated.

On 22 May 2009, the Minnesota Department of Health released a report ('Public Health Impacts of Wind Turbines'), detailing the health impacts from wind turbines and low-frequency noise vibrations. The conclusions indicated that wind turbines emit low-frequency sound which may affect people in their homes, especially at night. The report stated: 'The most common complaint in various studies of wind turbine effects on people is the impact on quality of life. Sleeplessness and headache are the most common health complaints and are highly correlated with annoyance complaints. Complaints are more likely when turbines are visible or when shadow flicker occurs.' The report concluded that: 'Shadow flicker can affect individuals outdoors as well as indoors and may be noticeable inside any building.'

A June 2009 report ('Sleep Disturbance and Wind Turbine Noise') by the British physician Dr Christopher Hanning stated that: 'Families whose homes were around 900 metres from wind turbines found the noise, sleep disturbance and ill health eventually drove them from their homes.' He went on to state: 'Inadequate sleep has been associated not just with fatigue, sleepiness and cognitive impairment, but also with an increased risk of obesity, impaired glucose tolerance (risk of diabetes), high blood pressure, heart disease, cancer and depression. Sleepy people have an increased risk of road traffic accidents.' Many of these symptoms are suffered by victims of wind turbines, such as Kay Siddell and Norrie Gibson, who were mentioned earlier in this book.

Even earlier, in 2006, the French Academy of Medicine issued a report that concluded: 'The harmful effects of sound related to wind turbines are insufficiently assessed. The sounds emitted by the blades are low frequency, which therefore travel easily and vary according to the wind, constituting a permanent risk for the people exposed to them.'

Most scientific surveys in this field have concluded that the only mitigation of adverse impacts from turbine noise or flicker effect is to construct the wind turbines a sufficient distance away from places of human and animal habitation. The safest distance generally mooted is between 1.5km and 5km, although some experts have suggested 10km is preferable. Dr Sarah Laurie, medical director of the Waubra Foundation, an Australian body dedicated to researching the health effects of wind turbines close to human habitation, argues that a 10km exclusion zone is necessary to avoid impacts which she claims can be 'life-threatening'.

Dr Laurie says that such precautionary setback zones can help avoid conditions which she has witnessed in patients who have suffered long-term exposure to wind turbine noise, vibration and flicker effect. The former GP says that she has examined patients who have experienced an adrenalin surge leading to severe headaches and dangerously high blood pressure. She has called for more research to establish what exactly such people have been exposed to and what their symptoms are.

Dr Chris Hanning, a retired consultant in sleep medicine, said Dr Laurie's recommendation for a 10km exclusion zone until such time as further research was conducted was sensible. 'This would certainly prevent any harm while further investigations are being carried out,' he said. Dr Hanning continued: 'Health impacts of wind farms are serious. I have no doubt that many people have suffered serious adverse effects. The Japanese Government has implemented a four-year programme of research into the health effects of wind turbine noise. Pressure should be placed on the UK and Scottish Governments to do likewise and in the meantime enact a moratorium on onshore wind farm construction' (*Express* 28/11/2011).

In Scotland there is currently a 2km recommended guideline for separating wind turbines from the edge of villages, towns or cities, but this is not mandatory and it is left up to planning authorities to decide on the distance that wind farms can be erected from people's homes. As a result, the 2km guideline is widely ignored, as we have seen in examples in this book. But despite all of this weight of evidence, the Scottish Government and the energy companies remain unconvinced. 'There is no credible evidence that wind farms affect health and no evidence of health effects arising from low-frequency noise from wind farms,' said a Scottish Government spokesperson (*Scottish Daily Mail* 20/08/2012). Niall Stuart, chief executive of Scottish Renewables concurs: 'We are not aware of any serious piece of scientific research which has shown wind turbines to have a negative impact on human health' (*Sunday Post* 19/08/2012).

Wind farms can also cause catastrophic accidents. Most people are familiar with the turbine that burst into flames and exploded at Ardrossan during a fierce storm in 2011, and many will have heard about the houses that were evacuated when a turbine had to be dismantled and came crashing to the ground in Berwickshire. But these two well-publicised events merely scrape the surface. There have been 1,500 other incidents involving UK wind farms over the past five years, which include four deaths and 300 injuries to workers. Alarmingly, there is an average of one incident per day, including cases where 14-tonne blades have crashed to the ground.

There was also a controversial case involving a turbine in the playground of a school on the island of Raasay which was shedding springs around the playground and had to be stopped by the head teacher with the assistance of a parent. It was later lowered to the ground for inspection, removed and not replaced. The inspection report stated: 'Ultimately the machine was running unloaded and ungoverned in high winds, which is a very noisy and dangerous state for a wind turbine. The position of the turbine directly outside a school maximised the danger to the public and to small children.'

Despite this, Highland and other councils in Scotland continued to install turbines within school grounds. Highland Council made several recommendations for the future, including an exclusion zone, most of which were ignored. Finally, in the spring of 2012, Highland Council was forced by public pressure to shut down all wind turbines erected in school playgrounds. When I raised this matter with the Scottish Government's Energy Minister Fergus Ewing in a face-to-face meeting in Brussels in early 2012, he said he was 'unaware' of any turbines having been built in school playgrounds or near schools anywhere in Scotland!

Impact on national security

The Scottish Government's policy of support for onshore and offshore wind energy must be considered as a matter of national security. With our armed forces already stretched to near breaking point and further cuts damaging the morale of our soldiers, sailors and airmen, it is absolutely incomprehensible that the MOD is forced to devote valuable resources to the sole purpose of ensuring that wind turbines do not threaten any national defence structures.

The Wind Energy Team at Defence Estates has to assess every wind farm on a case-by-case basis to ensure that turbines do not create hazards or interfere with defence installations. They have to evaluate every single cluster of turbines to decide if they create obstacles in Low Flying Areas. This is a serious problem in Scotland because military jets fly as low as 150ft in some exercises and there are several tactical military training areas, including the Highlands and South Lanarkshire, around the country where such low flying is commonplace.

By law, turbines taller than 300ft must be fitted with lights so that they are visible to pilots. But the Civil Aviation Authority has warned in the past that the lack of lights on smaller structures could cause planes to crash through the blades when flying low.

The MOD's Wind Energy Team has spent vast amounts of time and resources trying to rectify the problems that wind turbines cause for its radar equipment. Turbine blades are known to emit microwave radiation which often interferes with the primary radar of aircraft and secondary surveillance radar and navigation aids.

Worryingly, a build-up of wind farms can give the appearance of a 'moving object' which confuses air-traffic controllers into believing that an unidentified aircraft has entered the area. Turbine blades can rotate at up to 200mph, mimicking on screen the appearance of slow-moving aircraft. Needless to say, such a scenario could have a potentially disastrous effect if other aircraft are in the area. Furthermore, some observers have even warned that turbines will create blind spots in radar coverage known as 'blackout zones'. Such blind spots mean that hostile aircraft could evade detection.

Indeed, senior managers at Scotland's main international airports have complained to me that they now have to employ full-time members of staff simply to lodge objections to inappropriately sited wind farms which could impact dangerously on their radar and navigational systems. They are deeply frustrated that in an increasingly competitive aviation market, they are forced to divert a great deal of energy and resources into dealing with these egregious turbine applications, rather than concentrating on running their airports profitably. Again, this points to a failure of government and a lack of proper zoning, to ensure that comprehensive buffer zones were drawn around all of Scotland's main airports within which no wind turbine applications would be entertained.

With many new wind farms being built across Scotland's open countryside, the MOD is also increasingly concerned about the impacts on their radar installations. In June 2012, Buchan councillors rejected a proposal for a 20KW turbine at Honeyneuk Farm, near Maud, Buchan, after the MOD opposed the project due to concerns over radar interference at RAF Buchan. Glen App wind farm in Ayrshire has been questioned by the MOD as its 36 turbines, each rising to over 126 metres, will cause unacceptable interference with range-control radar at West Freugh RAF station, 17km from the site. The Wind Energy Team at Defence Estates is concerned that the turbines will create false targets and will confuse surveillance in the area. For interference to be negated, the developers would have to keep the turbines to a maximum of 52.5 metres high.

In late 2011, the MOD objected to the proposal to build two 132-metre turbines at the GlaxoSmithKline Montrose plant. The turbines would be sited 43.9km from the ATC radar at RAF Leuchars and, according to the Defence Infrastructure Organisation, 'will cause unacceptable interference'. In the objection letter sent to Angus Council, the MOD noted that turbines detrimentally affect MOD radars by causing 'the desensitisation of radar in the vicinity of the turbines, and the creation of "false" aircraft returns which air-traffic controllers must treat as real'. Fortunately, in September 2012, Angus Council announced that they had refused consent for this controversial proposal.

The MOD goes on to point out that 'maintaining situational awareness of all aircraft movements within the airspace is crucial to achieving a safe and efficient air-traffic service, and the integrity of radar data is central to this process. The creation of "false" aircraft displayed on the radar leads to an increased workload for both controllers and aircrews, and may have a significant operational impact. Furthermore, real aircraft returns can be obscured by the turbine's radar returns, making the tracking of conflicting unknown aircraft (the controllers' own traffic) much more difficult.'

Many other turbine proposals have been rejected across the UK, because of MOD objections. As such, energy companies, desperate to gain planning approval for new industrial wind farms, have even resorted to paying for wind-farm-friendly radar systems which they then give to the MOD in return for their withdrawal of objections. Wind farm developers have moved quickly to strike a pioneering deal whereby the wind farm industry will spend at least £16 million on advanced radar defence systems in order to pave the way for more turbines.

Since developers first began planning wind energy projects, they have often been delayed due to planning disagreements resulting from potential impacts on air-traffic control and defence systems. Thus, a consortium of wind energy giants is investing in US-built Lockheed Martin radar equipment to circumvent the problems caused by the turbines.

The Lockheed Martin Air Defence Radar TPS 77 system, which costs approximately £20m, will be paid for by developers, the Crown Estate and the Department of Energy and Climate Change. At the moment, the new radar systems are mainly

1 The 'secret valley' at High Tralorg Farm, where Kay and John Siddell's lives have been ruined by 52 giant industrial wind turbines. (Photo: Struan Stevenson)

2 The nearest giant turbines are only 742 metres away from the Siddells' home. (Photo: Struan Stevenson)

3 Whitelee windfarm (ScottishPower Renewables – owned by Spanish firm Iberdrola) is Europe's largest, with over 200 giant turbines built on deep peatland and on areas cleared of trees. (Photo: Struan Stevenson)

4 The roof of Norrie Gibson's house at High Myres, in the centre of the windfarm at Whitelee, south of Glasgow. (Photo: Struan Stevenson)

5 A gigantic turbine at Whitelee towers above a van. (Photo: Struan Stevenson)

6 Struan Stevenson standing at the foot of one of the giant turbines at Iberdrola's windfarm at Maranchón, Guadalajara, Spain. (Photo: Struan Stevenson)

7 Inside the tubular structure of an industrial wind turbine at Maranchón. (Photo: Struan Stevenson)

8 Professional visualisation of the wind array proposed for the Firth of Clyde, as seen from the Championship Golf Course at Turnberry, where the nearest turbine will be only 2.8 km from the shore. (Image: Malcolm Kirk)

9 *'The re-industrialisation of Scotland'* has become a reality for the inhabitants of Barrhill, South Ayrshire, who live next to the Arecleoch windfarm. (Photo: Struan Stevenson)

10 (Top) The Environmental Impact Assessment for the proposed Moray Firth offshore development was so large that it had to be delivered on a palate. At over 70,000 pages, MORL ensured that few will ever read the entire document to check that the turbines will not detrimentally impact on the marine environment. (Photo: Bertie Armstrong)

11 (Left) A road sign erected at the exit from ScottishPower Renewable's Mark Hill windfarm at Barrhill Ayrshire. It warns Spanish site workers to drive on the left! (Photo: Struan Stevenson)

12 Cartoon by Iain Green which appeared in the *Scotsman* on 10 November 2011, on the eve of the first CATS National Windfarm Conference in Ayr, when Struan Stevenson delivered a speech, entitled 'National Follies'.

13 Struan Stevenson after test-driving a Honda FCX Clarity hydrogen-powered car in 2010, outside the European Parliament, Strasbourg. (Photo: Struan Stevenson)

focused on the area known as the Wash, near Norfolk, in order to protect Britain's sensitive eastern airspace approaches. However, two more 'wind-farm-friendly radars' have already been ordered for sites at Staxton Wold, North Yorkshire, and Brizlee Wood, Northumberland, potentially unlocking a further 750MW of proposed projects.

Besides concerns about onshore radar installations, the MOD is also blocking plans for hundreds of turbines which will disrupt seismological equipment at the Eskdalemuir station near Lockerbie, which listens out for countries secretly testing nuclear weapons. As the UK is a signatory to the UN Non-Proliferation of Nuclear Weapons Treaty, it is required to maintain a seismic listening station to monitor any seismic vibrations caused by nuclear explosions. But wind turbines produce vibrations which travel through the ground and disrupt seismological instruments. Consequently, the MOD has stopped all wind farm developments within a distance of 10km from Eskdalemuir and they are restricting other developments within a radius of 50km from the station because they claim that the 'noise budget' within this zone has already been attained.

Whilst onshore turbines will affect the UK's ability to detect nuclear explosions around the world, offshore turbines will affect communications facilities and naval operations, particularly those of our Vanguard-class and Astute-class submarines based at Faslane. MOD sources have warned that offshore wind farms could hamper access to the Faslane base on the Clyde. The submarines access the base via the Firth of Clyde, which the Scottish Government has designated as a potential site for a massive offshore array. This strategy may, of course, be to the liking of Alex Salmond and the SNP, who wish to rid Scotland of the UK's nuclear deterrent.

Since coming to power, the Scottish Government has done little to ensure our nation's energy security, but risking our national security with an ill-conceived and obsessive energy policy is a step too far.

CHAPTER SEVEN

Are Offshore Wind Farms the Answer?

With public opposition to onshore wind power taking root, the focus is shifting to offshore turbines. The development of offshore wind energy in the UK has been slower than anticipated, as it has been more costly than experts originally predicted, but it is set to expand rapidly as the Government strives to meet renewable energy commitments.

At the European Wind Energy Association's offshore wind conference in Amsterdam in November 2011, Fergus Ewing announced that the Government had identified 15 areas where vast offshore arrays would be constructed. He chose the relative safety of Amsterdam to unveil a map showing how gigantic offshore turbines will imprison the entire coastline of Scotland. The plans represent the total industrialisation of Scotland's seascape, virtually enclosing and surrounding the country from Berwickshire up the east coast to Shetland and back down the west coast to the Solway, with massive offshore wind projects.

The 15 potential new offshore sites will destroy forever the beauty of the Berwickshire and East Lothian coastline, the Firth of Forth, the Moray Firth, the Northern Isles, the rugged coast of Sutherland, the Western Isles, the Firth of Clyde and the Solway Firth. A monstrous array of turbines stretches from Lochboisdale in South Uist to Tobermory in Mull, completely engulfing Tiree and Coll, stretching for more than 60 miles and appearing on the Marine Scotland plans as almost three times the size of the Outer Hebrides. Another huge array, twice the size of the island of Arran, almost closes the entrance to the Atlantic from the Firth of Clyde stretching from Kildonan to Campbeltown and halfway to the Antrim coast.

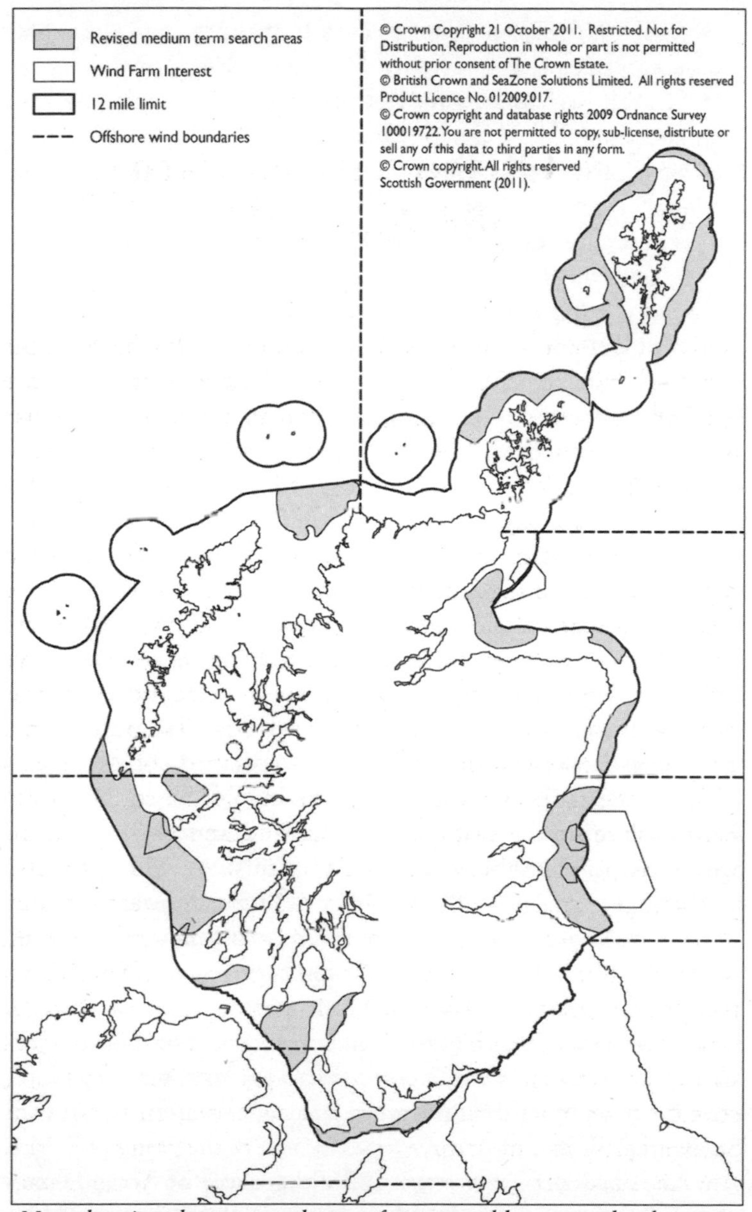

Map showing the proposed areas for renewable energy development around Scotland

By June 2010, eleven of the 'Round 1' offshore wind farms had been completed with a capacity of under 1GW. Approximately half the proposed capacity for Round 1 is either still being built or has been scrapped. Although Round 2 is not faring much better, the Department for Business, Enterprise and Regulatory Reform has launched Round 3 of UK offshore wind development. Round 3 is set in the context of the EU's 2020 targets and greatly exceeds the capacity of the first two rounds.

Round 3 ambitiously aims to triple the UK's current offshore capacity by 2020. The UK Energy Research Centre (UKERC) has suggested that this will require 15 to 20 GW of new offshore wind by 2020, but other observers have estimated that the UK must deliver up to 32GW of new offshore wind power by 2020, with much, much more to come in the following decades. This equates to some 6,400 turbines that need to be installed and grid connected in less than eight years. The scale and cost of such plans is unprecedented.

The Institution of Mechanical Engineers suggests that over £100 billion will be required to deliver this new capacity. Additionally, the cost of offshore turbines has been reported to have risen three-fold to £3 million per MW installed. This means that the industry relies heavily, if not completely, on the Renewables Obligation and other subsidies to meet its objectives.

In terms of area, offshore wind farms eclipse their onshore counterparts. The proposed Beatrice offshore wind farm being developed by SSE in the Moray Firth will have between 142 and 277 turbines, depending on turbine size, covering an enormous area of 131.5km^2. Two demonstration turbines already in position can be easily seen from many parts of the Caithness and east Sutherland coastline, so the visual impact of an offshore wind factory of this magnitude leaves little to the imagination. However, the prize for the biggest wind farm of all will go to Morl in the Moray Firth.

What is being billed as the world's largest offshore wind farm, with 339 turbines at a cost of £4.5 billion, will be constructed 12 miles off Caithness by Moray Offshore Renewables Ltd (MORL), a joint venture between Spanish/Portuguese firm EDP Renewables (EDPR) and Spanish oil and gas company Repsol Nuevas Energias. Work will commence in 2015 and be completed by 2020. MORL says that the turbines, which will tower 200 metres above sea level, will cover an area of 295km^2. They claim that they will produce up to 1,500MW of power. Dan Finch, project director and managing director of EDPR UK, estimates 'that the project will be capable of supplying the electricity needs of 800,000 to 1,000,000 households' (*Scotsman* 03/09/2012).

The project will have an estimated lifespan of around 20 years, during which time the turbines will require constant repair and maintenance due to the harsh conditions in which they have to operate. We will also have to rely on constant base-load back-up from coal- or gas-fired power stations to keep the lights burning on the days when there is no wind or when the wind is blowing so strongly that the Moray Firth turbines have to be shut down.

The cost of the project, set at £4.5 billion, is therefore simply not economically sustainable. A new state-of-the-art nuclear power station capable of producing 1,200 MW of power, operating at 80% efficiency and with an estimated lifespan of 60 years, extendable to 120 years, costs between £2.5 and £5 billion, including all decommissioning and waste disposal costs. And nuclear power is virtually CO_2 emission free.

The conclusion has to be that the SNP Scottish Government is determined to give this monstrous project and others like it the go-ahead because it fulfils the dogmatic prophecies of their 'green', anti-nuclear agenda. The vast subsidies for this and other offshore wind farms are simply passed straight down the line to the electricity consumers, leading to repeated hikes in our bills and, as we have seen, driving more than 900,000 Scottish households into actual fuel poverty. The project is not being built off the coast of Spain by this predominantly Spanish company because the Spanish Government has seen the light and ended all subsidies for wind turbines. There would be no onshore or offshore wind farms in Scotland either if there were no subsidies.

Rising fuel bills will drive industry out of Scotland and destroy jobs, just as happened in Spain. The devastating visual impact of this development will destroy tourism around Caithness and the Moray Firth. The pile driving and laying of concrete foundations over almost 300km^2 of the Moray Firth will have a catastrophic impact on marine ecosystems and sea mammals and the long-term noise and vibration from the 339 giant turbines will drive most sea life out of this formerly productive fishery. If anyone thinks this proposal is green they should think again.

The European Environment Agency warns that for any area up to 10km from the coast, the visual impact of wind turbines is significant. The Dutch Government has banned the construction of wind farms built within 22km (12 nautical miles) of the coast, due to their significant visual impact.

In the UK, the strategic environmental assessment for Round 3 of offshore development has proposed to give preference to locations beyond the 12-mile zone, but we know that with wind farms, 'recommendations' and 'preferences' are almost always ignored by developers and planners. We saw it with the recommended 2km separation distance between onshore turbines and residential homes in Scotland, which has been repeatedly ignored, and we will see it again with offshore wind farms.

Offshore wind turbines may also have a negative impact on the local environment. No one seems to know what effect so many turbines in a confined marine area could have, giving rise to serious concerns among the local communities involved. Of course, consultation with ordinary communities and businesses potentially affected by these life-changing proposals has been kept to a minimum. Fishermen in many parts of Scotland say their world-renowned shellfish grounds will be wrecked forever by the excavation and construction of turbines, not to mention the on-going vibration and disturbance which will drive fish out of our waters.

Sailors say that no one will wish to risk slaloming around these towering steel giants and the Royal Yachting Association has warned that turbines will not only cause visual intrusion but will also interfere with navigational equipment (RYA 2004). Tourists will treat Scotland as an industrial wasteland, best avoided. Coastal communities who have relied for generations on fishing and tourism will be destroyed. But like onshore wind, offshore wind is ill-conceived and unsustainable as a coherent energy policy. Compared to onshore wind power, offshore is more difficult to install, requires more maintenance, is more expensive and is even more heavily subsidised. It accounts for less than 2% of the global installed wind capacity and, as an infant technology, so-called 'first of a kind' costs still apply. So it is inexplicable that any government would seek to push forward with the kind of huge development plans being mooted for Scottish waters, the magnitude of which is unparalleled in the UK.

Sir David King, chief scientist and adviser to the Labour Government under Tony Blair, commenting on a report from the Smith School of Enterprise and the Environment (SSEE) at the University of Oxford, published in April 2012, said, 'If we went all the way attempting to provide the energy we need with offshore wind, I believe the costs to the UK taxpayer would be simply phenomenal. They [offshore wind turbines] are extremely expensive to install and expensive to maintain.'

Offshore wind power is a much less developed technology than onshore wind. Latest figures from the European Wind Energy Association show that a total of 1,503 offshore turbines are now installed and grid-connected in European waters, totalling 4,336MW spread across 56 wind farms in 10 countries. The offshore wind capacity installed by the end of 2011 produces, in a normal year, 14TWh of electricity, enough to cover just 0.4% of the EU's total consumption. The UK is by far the largest market with 2,362MW installed, representing over half of all installed offshore wind capacity in Europe. But in terms of actual energy output for offshore, we are still building the equivalent of the UK's first conventional power station.

Put simply, offshore wind is inordinately expensive and there is no economic case for it. It is the most expensive large-scale commercially available low-carbon generator in the UK. When North Hoyle wind farm in northern Wales became the first offshore array to start generating electricity in December 2003, there was a widespread assumption that costs would fall as deployment expanded. But the UK Energy Research Centre (UKERC) noted that the financial difficulties and cost escalations have been significantly larger compared to onshore wind. Whilst pro-wind advocates herald onshore wind as the lowest cost commercially available renewable technology, offshore is still by far the most expensive.

The early-cost data for offshore wind farms did support the theory that costs would decrease over time. The world's first offshore array, Vindeby wind farm in Denmark, cost approximately £1.82 million per MW in 1991. By 2002, the costs had been reduced to around £1.05 million per MW when Horns Rev wind farm was built off the Danish Coast. In 2003, North Hoyle was constructed at a cost of £1.35 million per MW and the following year Scroby Sands, near Great Yarmouth, cost £1.26 million per MW.

However, since the mid-2000s, the costs of offshore wind have escalated dramatically. The CIVITAS think tank recently reported that in the last five years costs have doubled from approximately £1.5 million per MW to over £3 million per MW in 2009. The UKERC found that for projects which were grid connected since 2008, the capital costs have doubled compared to 2003 levels because of cost increases for materials, commodities and labour. Currency movements, increasing turbine costs, increasing turbine depths, planning and consenting delays and supply chain constraints have also had a significant impact.

The future is also looking bleak. CIVITAS estimated that Department of Energy and Climate Change green policies could be adding 45% to electricity costs by 2030 for owners of medium-sized businesses. These extra costs will damage competitiveness and undermine viability, especially for high-energy users. They risk driving industry to migrate overseas along with their CO_2 emissions, thus having zero net impact on global emissions totals. We have already seen this happening in Spain.

The turbines themselves represent the largest single-cost item in an offshore wind farm, up to around half of overall capital expenditure. Offshore turbines are approximately 20% more expensive than their onshore counterparts as they must operate in a continuously hostile environment where reliability is imperative, given the difficulty of maintaining wind turbines at sea.

Offshore foundations are another exorbitant cost because they are more substantial and more complex to lay than onshore turbines. Offshore installation is a challenge which is normally limited to the summer months and it goes without saying that the more windy the site, the greater the risk. With the exception of the Beatrice development in the Moray Firth, no offshore foundations are manufactured in the UK. Most are sourced from Holland and are subject to varying steel prices and sterling–euro currency fluctuations.

Depth and distance from the shore are of particular relevance to future Scottish offshore wind developments. Most operational offshore turbines have been built in shallower water, usually in depths less than 25 metres, with either piled or gravity base foundations.

In a 2009 Technical Report on offshore wind, the European Environment Agency noted that offshore wind farms have not been built in waters with depths above 30 metres, but this will change in the future. Currently, monopile foundations are the foundation of choice for offshore turbines, but technological advances will soon allow the construction of taller and more expensive deep-water turbines which will utilise quadropod jacket bases in waters of around 45 metres depth. Further advances, such as tripods, jacket and floating structures, could eventually enable turbines to be situated in waters up to 100 metres deep. But nobody has determined if better wind speeds will compensate for the additional costs associated with going further offshore.

Furthermore, our docks and ports will need considerable investment if we are to install the turbines ourselves. A key problem facing Scotland is that our vessels and the wider installation supply chain are tightly constrained and the wind industry must often compete with offshore oil and gas. Whilst there are questions that financial experts must answer, it is clear that offshore wind power is expensive. Therefore it seems unwise, if not absolutely absurd, to be embarking on a huge programme of investment in offshore wind-generated electricity, especially when we already face grave economic challenges.

The Institution of Mechanical Engineers, who do support wind power in theory, see the key barriers to achieving 2020 targets through offshore wind as:

1. Inappropriate infrastructure
2. Inadequate technology
3. A deficiency in skills
4. Lack of manufacturing capability
5. Lack of funding.

Our electricity grid is outdated. It was built to connect large centralised electricity generating plant to industrial and domestic customers, not to facilitate remote power generators using local renewable sources. It needs a multi-billion-pound investment, which consumers would pay for. The Institution of Mechanical Engineers estimates that over 60% of the grid is outdated and must be upgraded in the next five to ten years. It would be a disaster if we were to fund all these costly changes to the grid, only for future governments to realise what a scam large-scale wind power is and pull the plug on the entire policy. Our governments must ensure transparency. They must tell us where the money is sourced, where it is spent and they must demonstrate best value for money.

Significant development work is needed to improve the efficiency of offshore wind technology. The Scottish Government claims that this will create jobs in Scotland, but we simply don't have sufficient numbers of qualified personnel for the development, assembly, operation or maintenance of this emerging offshore technology. Can Scotland really expect to train the necessary manpower, install the turbines and update the grid in just eight years? No. We will end up heavily subsidising the entire industry and creating huge numbers of jobs for the countries that can complete the tasks for us.

Although Scotland has world-class manufacturing industries, we lack a sufficient manufacturing base for the large volume of equipment which will be required to meet the 2020 energy targets. The Scottish Government has forgotten that for a target to be realistic it has to be founded on factual data and a comprehensive engineering-based technical assessment. To quote experts from the Institution of Mechanical Engineers, 'if a target is not achievable there is no point setting it' and right now there is no practical strategy in place to ensure that Scotland will achieve its Government's 2020 targets.

No comprehensive engineering assessment has been published in the public domain which would support the targets. Although they fully support the desire to maximise the enormous potential for renewable energy, from an engineering perspective, the Institution can't see how a sufficient installation rate will be achieved through current policies. They urge the government to refocus on a pragmatic, 'real-world' approach to what can actually be realised. They want the Scottish Government to state clearly its engineering-based methodology for achieving their ambitious targets without delay.

Impacts on marine life

It is not just the financial aspect of offshore turbines which causes concern; they will also seriously affect marine life, including many species which have provided a long-established livelihood for our fishermen. There is very little good-quality scientific evidence on the effects of offshore turbines on marine species and fisheries. But leading marine biologists have warned that claims about their potential for positive effects, such as enhancing the potential for the expansion of nursery grounds, are based on supposition rather than research, or they are extrapolations from small-scale studies of very specific situations which do not really support the generalisations on which they are based.

The Scottish Government's plans to surround our entire coastline with wind turbines will not only endanger Scotland's commercial fishing fleets, but will also jeopardise our recreational sea angling industry, which is worth over £140 million per year to the Scottish economy and supports in excess of 3,000 jobs. The Outer Solway region in Dumfries and Galloway is one of the most valuable recreational sea fishing grounds in Scotland. Although sea angling contributes over £23.79 million to the region every year, the Danish wind farm giant DONG Energy has still been allowed to examine the area's potential for a second wind farm beside the existing 60-turbine Robin Rigg development.

The UK Commission for Employment and Skills noted that the current Robin Rigg development, owned by E.ON, has an operations team of just 27 people, with 22 of those jobs being held by workers from the local area. The Commission states that, in conjunction with local authorities from Dumfries and Galloway and Cumbria, a community fund of £1 million was provided by the developers. The Port of Workington was chosen as the centre for operations and maintenance during the 20-year lifespan of the Robin Rigg development. It is estimated that the operation and maintenance tasks will inject £2.5 million into the local economy. So not only will the over-exploitation of our fragile marine resources harm both our inshore fish stocks and the local communities, but the supposed financial benefits of the wind farm will pale in comparison to the loss of revenue caused by the impacts on activities like recreational sea angling.

As previously mentioned, the hostile ocean environment necessitates offshore turbines having massive foundations with huge piles driven deep into the seabed. The resultant noise created causes sound pressure levels which seriously damage the hearing systems of marine mammals. Research has shown that cod and herring can detect this noise up to 4km away and dab and salmon can be affected from a distance of only 1km.

For acoustically sensitive animals such as harbour porpoises and harbour seals, which require their hearing for orientation, communication and survival, the zone of audibility for pile driving will most certainly extend beyond 80km, perhaps even hundreds of km, from the source. Once the turbines are constructed, operational noise is less powerful than pile-driving noise but it still has the potential to affect marine life behaviour over distances of several hundred metres from the pile. These statistics are even more worrying given that Scotland holds approximately 79% of the total UK population of harbour seals, a population which is already in decline on a country-wide scale.

The No Tiree Array campaign – resisting the proposed massive offshore wind farm which starts only three miles from the shores of Tiree and wraps around the island's south and west coasts – issued an information update at the end of August 2012 in which they highlighted information on the movements of basking sharks in the area.

According to Robert Trythall, Secretary of the No Tiree Array group, the developer in question, Scottish Power Renewables, had carried out a survey of basking sharks – a simple head count – and were reported to have sighted no fewer than 914 in a single day within the proposed area of the Argyll Array, now better known as the Tiree Array. Whatever the accurate numbers, there can be no doubt that these waters play host to a great abundance of fish, sea mammals and cetaceans, all of which will be severely impacted by offshore wind developments such as the Argyll (or Tiree) Array.

Detrimental impacts on marine life will have knock-on consequences for our fishermen. A good example of such impacts can be found in East Yorkshire where Mike Cohen, the CEO of the Holderness Coast Fishing Industry Group, who also happens to be a marine biologist, has provided detailed evidence of how offshore wind farms can cause serious and permanent damage to marine ecosystems.

He says Holderness is home to a vibrant, centuries-old fishing community, where its cold water means it is world-renowned for its high quality crab and lobster. Holderness is now one of the largest crab and lobster fisheries in Europe. The Holderness coastal fleet encompasses around 70 boats equipped with static fishing gear (lobster pots). They are mostly less than 12 metres and most are owned and operated by their skippers as independent businesses. This fleet is no different to many of the small-scale vessels which fish in Scottish waters.

Consent has recently been given for the construction of a $35km^2$ wind farm at Westermost Rough, one of the most important parts of the Holderness coastal fishery, which is one of the most productive lobster fishing grounds in European waters and is a major source of fishermen's income in winter. The rocky substrate, with plenty of small boulders and crevices, provides a near-perfect, but highly sensitive habitat for lobsters. The developer who has been given consent for the wind farm is DONG Energy – a Danish energy multinational with a growing interest in the UK's heavily subsidised renewables sector. DONG Energy has been working hard to bring offshore arrays to the Outer Solway and Duddon Sands in Scotland.

There are two key problems with the proposed wind farm at Holderness. Firstly, drilling into the seabed to provide foundations for the planned 80 turbines would cause massive environmental destruction by releasing large amounts of sediment into the water column. Upon settling, it would smother life on the sea bed, eliminating fishing in the short term. If sedimentation is deep enough to bury the habitat features that crab and lobster need to colonise an area, stocks in those areas would not return and fishing would be excluded indefinitely. There are already 'dead spots' like this which occurred after the construction of a nearby gas pipeline along the Holderness Coast.

The second problem, which concerns the effective exclusion of fishing vessels from the wind farm area, is particularly astounding. DONG's representatives have informed fishermen that they will still be 'allowed' into the site after construction, provided their activities are not seen as detrimental to the wind farm. *Allowed?* For centuries, UK vessels have had freedom of movement throughout UK territorial waters. Now it would seem that fishermen whose families have worked an area for generations need the permission of a Danish energy company executive to enter an area of sea that is just off the shore.

This is simply ludicrous. Normally, fishing vessels are not allowed within 500 metres of an offshore turbine. With up to 80 turbines in the Westermost Rough area, this will close off most of the site and render the rest impractical to fish. Static gear drifts with the tide, so if any were to drift within the exclusion zone around each turbine, it would have to be abandoned. Despite the 'permission' graciously granted by DONG's project managers, the construction of this wind farm equates to the de facto closure of one of Europe's best lobster fisheries.

Keeping Holderness in mind, it is galling to think what will happen to Scotland's fishermen and our world-class fishing grounds if the SNP Government gets its way and encases our coastline in turbines. Combine this with the government's recent plans to fast-track development plans for offshore wind farms and we have a serious situation. The Scottish Government needs to stop and take stock of its unsustainable renewables strategy, not accelerate delivery of a process that is seeing our world-renowned seascapes transformed into vast, rusting electricity factories that destroy tourism, fisheries and marine habitats, while driving Scottish households relentlessly into fuel poverty.

Those who claim that offshore wind arrays can provide a good habitat for marine organisms are deliberately trying to mislead us. Marine biologists have warned that this claim is based on a scientific study of fish aggregation around wind turbines, which also speculated that, in areas where wind turbines need rocks placed around their bases to protect against scour from suspended sandy sediment, this scour protection – NOT the turbines themselves – MIGHT prove a suitable habitat for crabs and lobsters in areas where no such habitat previously existed. Note that this was simply speculation in the conclusion of a paper describing related research. The *actual* effects on crab and lobster populations have not been studied.

The idea that drilling huge boreholes in an undisturbed and enormously productive habitat, inserting wind turbines into them and then dumping massive loads of rock around the base of each will somehow increase the populations of living organisms is utterly risible. It is little different to claiming that putting 80 wind turbines in the middle of a pristine forest will increase populations of owls and badgers! Offshore wind farms will damage the marine environment, a thriving traditional industry and the livelihoods of hundreds of ordinary people; and all of this in pursuit of a technology that is unreliable and inefficient, so that the SNP Government can fulfil its obsession with wind power.

CHAPTER EIGHT

Impacts on Marine Ecosystems, Forests and Peat Bogs

Impact on marine ecosystems

As shown in chapter seven, offshore wind farms will not only fail to provide national security, they will also have serious impacts on marine animals and the communities which rely on them. However, it is not only fishermen who will suffer. Few people realise that oceans play a significant role in the global carbon cycle.

Oceans are the largest long-term natural sink for carbon and they store and redistribute CO_2. Some 93% of the earth's CO_2 is stored and cycled through the oceans. Every day, we add a further 22 million metric tonnes of CO_2 to our oceans. That is why maintaining and improving the ability of our oceans to capture and store CO_2 is of vital importance to human survival. We cannot afford any longer to overlook the critical role of our oceans. Without the essential ecosystem service they provide, climate change would be far worse.

Recent research has indicated that a tiny part of the marine environment – the mangrove swamps, salt marshes and seagrasses that cover just 0.5% of the seabed – account for the capture of at least half, and maybe three-quarters of this 'blue carbon'. They are our blue carbon sinks. Keeping them in good shape could be one of the most important things that we do to keep climate change under control. Blue carbon sinks and estuaries capture and store the equivalent of up to half of the emissions from the entire global transport sector, every year. By preventing the further loss and degradation of these ecosystems and helping their recovery, we can contribute to offsetting 3–7% of current fossil fuel emissions within the next two decades – over half of that projected for reducing rainforest deforestation. The effect would be equivalent to at least 10% of the reductions needed to keep concentrations of CO_2 in the atmosphere below 450ppm.

Whilst mangrove swamps are found mainly in the tropics and subtropics, the UK possesses large areas of the other blue carbon stores, particularly seagrass meadows, kelp forests and salt marshes. Scottish National Heritage has noted that the vast majority of the UK's seagrass meadows and kelp forests are located in Scottish waters. In fact, within north-west Europe approximately 20% of seagrass beds are found within Scottish waters.

Conservation International and IUCN – The International Union for Conservation and Nature – have both highlighted that seagrasses around Scotland mostly occur in shallow coastal waters, probably at less than five metres depth, because of the relatively low water clarity of oceans at higher latitudes. Seagrasses need sunlight for photosynthesis. However, they are also commonly found at depths of up to 20 metres in clearer waters. Scotland has a long coastline, representing about 8% of the coastline of Europe, and consequently the ecosystem services provided by blue carbon stores around Scotland's coast are of global significance.

The rate of loss of these marine ecosystems is much higher than any other ecosystem on the planet – in some instances up to four times that of rainforests. Halting degradation and restoring both the lost marine carbon sinks in the oceans and slowing deforestation of the tropical forests on land could result in mitigating emissions by up to 25%.

Brown (light-absorbing organic matter and gases other than soot) and black carbon (soot) emissions from fossil fuels, biofuels and wood burning are major contributors to global warming.

Black carbon is thought to be the second largest contributor after brown carbon. Thus, reducing black carbon emissions represents one of the most efficient ways for mitigating global warming. Black carbon enters the ocean through aerosol and river deposition. It can comprise up to 30% of the sedimentary organic carbon in some areas of the deep sea and, according to scientists, may be responsible for 25% of observed global warming over the past century.

Green carbon is also a vital part of the global carbon cycle. It is removed by photosynthesis and stored in the plants and soil of natural ecosystems. So far, however, it has mainly been considered in the climate debate in terrestrial ecosystems, though the issue of marine carbon sequestration has been known for at least 30 years.

But while the loss of green carbon ecosystems (forests) has attracted much interest, the fact that nearly 55% of all green carbon is captured by living organisms not on land, but in oceans, has been widely ignored, possibly our greatest deficit in mitigating climate change. The carbon captured by marine organisms is called blue carbon. Blue carbon is the carbon captured by the world's oceans and represents more than 55% of green carbon. It is captured by living organisms in oceans and is stored in the form of sediments from mangroves, salt marshes and seagrasses. It does not remain stored for decades or centuries (like, for example, rainforests), but rather for millennia.

A recent assessment (Waycott et al., 2009) indicates that about one-third of the global seagrass area has already been lost, and that these losses are accelerating from less than 0.9% per year in the 1970s to more than 7% per year since 2000. About 25% of the area originally covered by salt marshes has been globally lost, with current loss rates at about 1–2% per year. About one-third of the area covered by blue carbon sinks has been lost already and the rest is severely threatened.

Marine vegetated habitats, blue carbon sinks, rank amongst the most threatened habitats in the biosphere, with global loss rates 2–15 times faster than that of tropical forests (Achard et al.,

2002). The loss of blue carbon sinks represents, in addition to the impacts on biodiversity and coastal protection involved, the loss of a natural carbon sink, eroding the capacity of the biosphere to remove anthropogenic CO_2 emissions.

Blue carbon experts wrongly assumed that offshore wind farm activities wouldn't disturb these systems as they believed that wind turbines are not usually built in shallow depths or near sandbanks where seagrasses often grow. But this is not the case. In the North Sea alone, there are already operational wind farms that have been built in shallow waters. In England, the Lynn and Inner Dowsing wind farm off the Lincolnshire coast has 54 turbines with a depth range of 6–11 metres. The Gunfleet Sands wind farm off Clacton-on-Sea, Essex, has 48 turbines in a depth range of 2–5 metres. Kentish Flats wind farm has 30 turbines in a depth range of 3–5 metres and Scroby Sands wind farm off the coast of Great Yarmouth in East Anglia has 30 turbines in a depth range of 0–8 metres. Many if not most of Marine Scotland's planned offshore developments will be in water depths of around 10–25 metres and will therefore have a heavy impact on seagrass meadows and kelp forests.

Offshore wind farms will also affect salt marshes – another invaluable blue carbon store. Salt marshes are vegetated parts of the upper intertidal area found on our enclosed shores. Scottish salt marshes make up 15% of the total British resource, and the largest areas of salt marsh in Scotland are located in two main areas: the Solway Firth and the Moray Firth. Despite the importance of these salt marshes, both the Solway and Moray Firths are key locations for wind farms.

The Solway Firth is currently home to the Robin Rigg wind farm, Scotland's first offshore wind farm, which was completed in April 2010. The Solway Firth houses 100 turbines in a depth range of 4–23 metres within a 59km^2 area, even though Scottish Ministers agreed that 'the Solway Firth and Wigtown Bay sites are unsuitable for the development of offshore wind and should not be progressed as part of this Sectoral Marine Plan' (Marine Scotland 2011).

Now a number of developers have revealed proposals to site huge wind farms in the Moray Firth Development Area as the Scottish Government has noted that the Moray Firth region 'has favourable conditions and significant potential for the development of offshore wind both within Scottish Territorial Waters and beyond into Scottish Offshore Waters'. The Scottish Government does not mention either salt marshes or seagrasses in their 'Sectoral Marine Plan for Offshore Wind Energy in Scottish Territorial Waters' (2011).

There are four European species of seagrass, two of which grow in Scottish waters: *Zostera marina* (eelgrass) and *Zostera noltii* (dwarf eelgrass). Globally, the estimated loss of seagrass from direct and indirect human impacts amounts to 33,000 km^2, or 18% of the documented seagrass area, over the last two decades.

Maerl* is very slow-growing and grows as unattached twig-like nodules on the sea bed. Maerl beds are home to myriad species including sponges, sea squirts, crabs, squat lobsters and clams. They provide feeding areas for juvenile fish such as cod and act as important nursery areas for commercially important species such as scallops, razor shells and edible crabs. Seagrass meadows are found around sandy coastal areas in water up to 20 metres deep. They provide shelter and living space for a wide variety of species, just like a meadow on land.

Destroying these habitats would release huge amounts of CO_2 into the atmosphere and would only add to climate change. If we are to tackle climate change and make a transition to a resource-efficient Green Economy, we need to recognise the role and contribution of all the natural carbon storage ecosystems. Halting degradation and restoring both the lost marine carbon sinks and slowing deforestation on land could result in mitigating emissions by up to 25%.

But to achieve this we have to reduce the rate of marine and coastal ecosystem degradation. We need to extend the emissions trading system to embrace blue carbon. Carbon credits for marine

* A collective term for several species of seaweed. It is a hard, chalk-like organism which grows in nodules.

and coastal ecosystem CO_2 capture and storage should be traded and dealt with in a similar way to green carbon. We need to establish a global blue carbon fund to pay for the protection and enhancement of remaining seagrass meadows, salt marshes and mangrove forests through effective management.

We need to improve energy efficiency in marine transport, including the fisheries, aquaculture and maritime tourism sectors. We need to encourage sustainable, environmentally sound, ocean-based energy production, including algae and seaweed. Vast industrial schemes to cover swathes of our coasts and ocean floors with wave, tidal and offshore wind farms must not be allowed to damage further marine ecosystem services.

Climate change is one of the drivers preventing the oceans from fully playing their role. One of the examples is acidification, which undermines their ability to buffer CO_2 emissions – cold waters having an enhanced absorption capacity – and jeopardizes marine biodiversity. As a result, if current trends in greenhouse gas emissions continue, many coral reefs may be lost over the next 20–40 years, resulting in devastating consequences for the numerous species they contain and for coastal communities around the world that would be less protected against extreme weather conditions.

Moreover, global warming leads to a proliferation of invasive alien species and to substantial modifications in the distribution and abundance of fisheries resources. Fishermen need to adapt to this new situation, but the socio-economic consequences arising from these changes are deeply worrying.

Despite the vital role of marine and coastal ecosystems in mitigating climate change, international and European policy makers pay too little attention to blue carbon. It is almost an unknown and too often overshadowed issue in the European decision-making process. For example the resolution of the European Parliament on the climate change conference in Durban (Nov/Dec 2011) recognised the importance of addressing deforestation and forest degradation. But there was no reference to the oceans.

Oceans are resilient but this treasure is at increasing risk.

Preserving and restoring coastal and marine ecosystems should be fully integrated into all climate-change mitigation strategies and biodiversity policies at international and European level. Sadly, not only is this not happening, in Scotland's case the reverse is true. Plans for offshore wind, wave and tidal energy projects threaten to wreak havoc in our coastal ecosystems.

Antoine Blondin, a famous French author, wrote: 'Bottles into the sea do not often bring back the answers.' The Scottish Government should stop searching for messages in bottles to answer their obsession with wind. They claim that they are seeking to reduce CO_2 emissions to combat climate change, but in fact they are destroying the planet while trying to save it! Worryingly, this is not the first time wind turbines have been built under the false pretence of reducing CO_2 emissions.

Impact on forests and peat bogs

The Forestry Commission's Annual Report in 2011 had an interesting statistic hidden away in the small print which revealed that in the past decade they have cut down between 12,000 and 25,000 acres of forest in Scotland to make way for wind farms. That equates to around 25 million trees and this at a time when government targets are calling for 25% of our land area in Scotland to be covered by trees, requiring the planting of 25,000 acres of trees every year. The Forestry Commission Scotland (FCS) has now granted exclusive rights to E.ON and ScottishPower Renewables to search all of its forests in Argyll for suitable sites for industrial wind turbine developments. In a self-confessed attempt to maximise income, FCS has invited expressions of interest from turbine developers for virtually any part of the National Forest Estate which covers 10% of Scotland's landmass. They openly declare their desire to make an income of £20 million per year from renewables by 2020. The FCS website blithely swats away any concerns about digging up peatland or cutting down trees on the National Forest Estate by saying all such projects will be subject to full environmental impact assessments.

It seems strange that following the criticism of Caroline Spelman's failed attempts to sell off England's forests, the SNP Government should now seek to turn over large swathes of FCS forestry to be cut down to make way for wind farms! Nearly 60 square miles of forests in Scotland have already been felled for this purpose. The Government plans to increase this area to 240 square miles or around 10% of the total FCS estate. The plans were revealed by SNP Finance Secretary John Swinney in his budget speech to the Scottish Parliament on 20 September 2012, when he spoke about 'the next phase of renewables developments on the National Forest Estate' (Scottish Government 20/09/2012).

So while we argue about costly new technologies to capture and store carbon in depleted undersea oil and gas wells, we crazily cut down vast swathes of forest every year, destroying nature's own carbon capture and storage system.

Worse still, we are digging up peat bogs all across Scotland to construct industrial wind developments. Peatland is Europe's equivalent of rainforest and it constitutes a vital component of the world's natural air-conditioning system. Peatland and wetland ecosystems accumulate plant material and rotting trees under saturated conditions to form layers of peat soil up to 20 metres thick – storing on average ten times more carbon per hectare than other ecosystems. But vast areas of carbon-capturing peat bogs in Scotland are being torn up to make way for so-called 'green' energy projects like wind farms, rendering the whole process carbon-negative.

Many giant wind turbines in Scotland are being built on deep peatland, causing immense damage to the environment and releasing vast quantities of CO_2. Hundreds of applications are still in the planning pipeline, many of them in wholly inappropriate locations which would threaten endangered flora and fauna and industrialise some of Scotland's most spectacular landscape. Worse still, by destroying deep peatland, these wind farms would create more carbon emissions than they would ever save.

Peat is a global carbon sink, storing millions of tonnes of CO_2 during the thousands of years the peat is formed from rotting trees and plant material. Damage to peat can extend as much as 250 metres on either side of turbine foundations and access-road installations. So the peat will gradually dry out over the years, resulting in an ongoing release of carbon. This can easily be calculated once the total extent of the planned development is known, using the fact that peat contains 55kg carbon/cubic metre – three times as much as a tropical rainforest! The whole hydrology of the area will change forever and, once damaged, peat can never be replaced – a terrible legacy to leave to future generations and a loss of a critical carbon sink.

Of course the big power companies, who are pocketing millions of pounds from these profligate projects, are keen to disprove this theory and regularly trot out 'experts' to say that drainage of the peat is not necessary and that damage to the environment will always be minimised. The 'floating roads' which are made to sound as if they can defy gravity by floating over the surface of peat bogs, actually require tens of thousands of tonnes of rock foundation, which cuts off the water flow to the bog and causes the peat to dry out, releasing millions of years of stored CO_2 into the atmosphere.

To suggest that a wind farm can be built without damaging peatland is just pure nonsense. As soon as the so-called 'floating' roads have been built and construction of the giant turbines takes place, the peat will be breached and drainage of the peat bog will occur naturally. This is basic hydrology! Drains will then have to be installed to take excess water off the site, otherwise the area will flood. This is called peat run-off and it will continually flow into adjacent watercourses, potentially causing the deaths of many freshwater and marine organisms as a result of suffocation. In the past, wind farm developers tried to mitigate peat run-off by filtering the water through straw bales. Such attempts were a total failure and, if anything, exacerbated the problem.

Taken together with the construction of the mammoth steel towers, vast concrete foundations under every turbine, borrow pits, drains, connecting and access roads, overhead power lines and pylons, it is not unreasonable to think that the carbon footprint from every wind farm built on deep peat far exceeds any environmental savings it may aspire to. As Lloyd Austin, the RSPB's Head of Conservation Policy, put it, 'there is no point building renewables that potentially emit more carbon due to peatland impacts than they save' (*Press & Journal* 25/07/09).

And the environmental destruction associated with wind farms doesn't stop there. Wind turbines require powerful magnets as a key component in their generators. These magnets are made from neodymium, which is extracted from rare earth metals in an industrial process that creates huge toxic wastelands in parts of China, poisoning land, crops, animals and people. So much for the so-called 'green' credentials of renewable wind developments.

The decision to refuse approval for the original application for a giant wind farm on peatland in Lewis was certainly a hopeful sign, although as is often the case, a new application was quickly lodged and approved for a £180 million wind farm at Eishken. Muaitheabhal wind farm on the Eishken Estate is a consented wind development on the Isle of Lewis. It will be sited on some of Scotland's wildest land and almost all of the turbines will be built on deep peatland. When completed, the wind farm will incorporate over 18 kilometres of access roads and 39 145-metre high turbines, each of which will be taller than the London Eye. The Isle of Lewis is not the only victim, as the Scottish Government intends to give consent to the construction of numerous large wind farms on the Western Isles, so that OFGEM will give their approval to the construction of a sub-sea cable to the mainland.

A massive onshore wind farm has been given the go-ahead in Shetland. The Viking wind farm being constructed by SSE Viking – a subsidiary of Scottish and Southern Energy – will have 103 turbines, making it the third biggest in Scotland. Once again the intention is to build the wind farm on deep peat with Scottish Energy Minister Fergus Ewing absurdly claiming that 'the development includes an extensive habitat management plan covering round 12,800 acres, which will restore peatland and offer benefits to a whole range of species and habitats' (Scottish Government

04/04/2012). This is like saying that cutting down tropical rainforest will provide wonderful grazing for cattle and sheep!

In a classic piece of 'double-speak' Stewart Stevenson MSP (no relation, although Royal Mail occasionally deliver his letters to me!), the Minister for Environment and Climate Change, stated in the foreword to SNH's 'Climate Change and Nature in Scotland' booklet that he welcomes Scottish Natural Heritage's ideas on peatland. He stated that 'Scotland has 15% of the world's blanket bog. Even a small proportion of the carbon stored in peatlands, if lost by erosion and drainage, could add significantly to our greenhouse gas emissions. Their importance to us on a global scale demands we should do more to protect this carbon store and to reduce emissions from degraded peatlands.' The duplicity of this statement is breathtaking. How can a Minister publicly acknowledge the global importance of Scotland's peat bogs whilst at the same time paving the way for wind farms to be built on top of them?

On the mainland, Dava Moor (Grantown on Spey), Gordonbush (Sutherland), Edinbane (Skye), Kergord Valley (Shetland) and many other wind farms have devastated or will devastate vast areas of peat bog. They should all be stopped. These developments constitute a potential technical breach of the Kyoto Protocol, which has a requirement that signatories protect their natural carbon sinks. Britain is a signatory to Kyoto.

Peatlands occur in 180 countries and cover 400 million hectares or 3% of the world's surface. Scotland has a unique role to play in preserving and maintaining this global resource. Over one sixth of the world's blanket bog is located in Scotland. Taken together with our forestry and moorland, the contribution Scotland makes to global carbon storage is significant.

Our peat bogs provide a unique resource which may shortly be recognised financially when a new carbon offset trading system is introduced. Landowners in Scotland should beware. If they cut down their forests or dig up their peat bogs to make way for wind farm developments, they may be throwing away a fortune. Owners of forests, peat bogs and moorland in Scotland could be looking at huge 'windfall' cash injections if new schemes for carbon trading take off in Europe.

Carbon trading works by company emissions being capped at a level that is equal to or below their previous levels. These companies are then given the opportunity to buy or sell carbon credits, depending on how much CO_2 they emit. The carbon credits can come from overseas. For traders, carbon dioxide is considered to be the world's newest commodity.

Forest owners in California are planning to sell carbon credits to industrial groups and power companies in exchange for deals to preserve redwoods and other trees which can soak up large quantities of greenhouse gases. This is a clear sign that owners of forests have woken up to the financial potential of offering carbon offsets to factories and energy companies in California where the state authorities plan to introduce a cap for emissions, allowing companies to set up a cap-and-trade system.

Investors in the US have estimated there is a significant potential for a multi-billion-dollar carbon market. In 2008 the global market was worth £75 billion. The US Congress is even discussing a federal plan that would roll out the scheme across the United States. When this happens, Europe is sure to follow suit.

At the Copenhagen Climate Change Conference in 2009, many of the participating nations spoke about the need to plant more trees to help in the natural capture and storage of CO_2. But a cap-and-trade scheme, which allows owners of major natural carbon storage systems like forests, peat bogs and moorland to trade carbon offsets with big industrial companies, would create a revolution in Scotland's more remote rural and upland areas. If Scotland's landowners could profitably develop these historically loss-making assets, there might be a chance that the unique landscape, wildlife, river systems and tourist economy could be saved from degradation for future generations. This would stop the mad dash by landowners in Scotland to dig up peat bogs and cut down trees in order to allow the construction of giant wind turbines, so that they can cash in on the significant subsidies from the production of renewable energy.

CHAPTER NINE

The Myth of 'Green Jobs'

Alex Salmond boasts that his obsession with renewables will create 130,000 'green' jobs. But in fact, with gas and electricity prices surging, jobs are being lost, more and more people are being forced into fuel poverty, and we are not even denting our carbon-emission targets.

According to KPMG, shifting from wind turbines to nuclear and gas-fired power stations would slash the UK's energy bills by £34 billion – or a saving of £550 for every person in the country (KPMG 2011). At a time when fuel bills have doubled in the past five years and are set to double again in the next five, savings like this are essential.

In addition, while the Scottish Government trumpets that greenhouse gas emissions have fallen to 57 million tonnes, down 13%, a recent report from the respected Stockholm Environment Institute (SEI) estimates Scotland's real carbon footprint has actually increased by 11% to 85 million tonnes (SEI 2009). So greenhouse gases are on the rise in Scotland. The SEI takes into account the greenhouse gas emissions created abroad from manufacturing things like massive steel wind turbine towers and then shipping them to Scotland for erection. The Scottish Government tries to hide these figures, disclosing only emissions within Scotland itself.

When you scrape the surface a little further you unearth many more discrepancies and double standards in Mr Salmond's assertions. For instance, he says that Scotland will have a world-beating low-carbon economy, and then in the next breath he applauds the fact that we have discovered huge new oil reserves in the North Sea that will keep us going until at least 2050!

He promises that Scotland will be nuclear free, while he knows full well that his policy of 100% energy from unreliable renewables will mean that we frequently have to import nuclear-generated electricity from south of the border where they are hoping to build eight new nuclear plants. And he claims Scotland will net big profits from selling electricity to England, but why on earth would England want to buy high-cost wind power from Scotland when they have just built eight new efficient nuclear plants of their own? It is an empty dream.

And Alex Salmond promises 130,000 jobs in his low-carbon economy with up to 50,000 to 60,000 alone coming from wind farm developments. In fact, the soaring cost of electricity, directly attributable to the race for renewables, is driving more and more businesses and industries to the wall. The number of companies going bankrupt in Scotland increased by 17% in the first quarter of 2012, while the number of individuals declaring themselves bankrupt rocketed by an astonishing 25% over the same period.

Indeed a recent study by the Whitehall-based Department for Business, Innovation and Skills (BIS) shows that soaring green energy charges 'will make British industry uncompetitive compared with other leading countries by the end of the decade'. (*The Times* 13/07/2012). The report states that a combination of renewable energy subsidies and new emissions charges will double the costs for energy-intensive industries, such as steel, by 2020, forcing many firms either to relocate abroad or close down altogether (BIS 2012).

The Scottish Government has chosen to ignore this report and is charging ahead with its plans to cover Scotland with wind turbines. But even this rash of wind farms is not creating Scottish jobs. If you visit any of the wind farm erection sites in Scotland you will hear a lot of foreign languages being spoken by the workers. And no wonder: the majority of the steel towers are delivered from Germany and Denmark on left-hand-drive trucks. The site workers are foreign, many from Spanish companies like Iberdrola and Gamesa. Many of the turbine blades are built in Scandinavia and many of the gearboxes in China.

In fact wind farm companies have taken on only a tiny handful of apprentices in Scotland. Claims that wind turbines would create thousands of jobs, particularly for young people, were exploded when it was revealed in August 2012 that only 20 modern apprenticeships have been created by the renewables sector out of a total of 26,427 apprentices who found employment. This represents a paltry 0.1% of the total (*Sunday Post* 26/08/2012).

At a meeting in Denmark on Wednesday 27 June 2012, Jens Martin Alsbirk – Director, Group Government Relations of the Danish renewables giant Vestas – told MEPs from the European Conservatives and Reformists Group (ECR) that his company started making wind turbines in 1979 and now have installed two thirds of global capacity – around 49,000 turbines. He said that Vestas commanded 12.7% of global market share for turbines in 2011, competing with companies like Siemens (Germany), Gamesa (Spain), General Electric (USA) and some Chinese companies. He said that although the sector had been hit badly by the financial crisis, Vestas was now building blades, towers, nacelles and generators in factories around the world. He boasted that his company accounted for 8% of total Danish exports and had created thousands of jobs in Denmark.

Meanwhile Germany announced in August 2012 that it intended to build 23 new coal-fired power plants, so that it could continue to offer energy prices that were competitive with other industrialised nations, according to the German Environment Minister, Peter Altmaier of the Conservative Christian Democrat Party. 'Green jobs' in Germany clearly involves companies like Siemens building wind turbine nacelles and other component parts to ship to the UK, while burning coal and pumping CO_2 into the atmosphere.

Indeed, the Scottish Government's claims that tens of thousands of new 'green' jobs could be created by 2020 may well turn out to be true. The sad reality is that most of these jobs will be created in countries like Denmark and Germany who manufacture the turbines. A leading consultancy, Verso Economics, found that for every green job created in the UK, 3.7 jobs are lost. A similar situation is happening across Europe. Thus, wind power is increasing energy costs and forcing energy-intensive companies to leave the country. No amount of subsidies will solve this because wind power just isn't competitive, it isn't affordable and it isn't reliable.

Even Vestas has felt the cold wind of reality. In June 2012 they announced that they were abandoning plans to build a factory in Sheerness in southeast England, where they had intended to manufacture items for their V164-7.0 MW turbines, causing shares in the company to fall by 2.8% on the Copenhagen stock exchange. So the likelihood of green jobs in the UK has been dealt another blow. Vestas also announced the closure of a major plant in China in June 2012, where they manufactured small kilowatt turbines, resulting in 350 job losses.

Richard Marsh, a renowned renewable power economist and Director of 4-Consulting, has slammed Alex Salmond's 'greatly exaggerated' claims of green job creation. Marsh noted that the widespread construction of turbines will not create 50,000 jobs by 2020, as envisioned by the SNP. In fact, the real number of long-term posts is more likely to be 1,100 but could be as low as 300.

Marsh was one of the experts who gave evidence during the renewables inquiry hosted by Holyrood's Economy, Energy and Tourism Committee in 2012. Marsh stated that the considerable difference between the SNP's forecasts and realistic figures is because the higher figure includes temporary construction jobs created while green energy projects are being installed. However, these jobs will cease to exist after construction is complete.

Marsh concluded that there are only around 1,100 green energy jobs in Scotland at the moment and the SNP's estimate is misleading and wholly dishonest because most of the jobs cited will be generated through the installation of infrastructure. Marsh finished by stating that these figures simply do not 'provide a compelling reason' for a taxpayer subsidy.

In May 2012, at a meeting of the Economy, Energy and

Tourism Committee, Citigroup's senior utilities analyst Peter Atherton said that the argument that the roll-out of renewable energy would provide a huge jobs boost in Scotland simply doesn't hold water. He said that only a small proportion of the billions of pounds spent on Scotland's renewable energy drive would reach the local economy and create jobs. 'Wind farms are capital-intensive operations that require relatively few people to build and operate them, therefore they are not great investments if the purpose is job creation,' he stated.

Atherton went on to explain how only 10% of the construction cost of the latest offshore wind farms is being spent in the UK, with 90% going to foreign suppliers, because most of the turbines being installed in Britain are being constructed in Germany, Spain and Denmark. The Citigroup analyst explained that because several of the UK's leading renewable energy companies are foreign-owned, and the fact that money to build factories and infrastructure in the UK will all have to be borrowed, will mean that a large chunk of the revenues from the industry will be siphoned off by international investors.

Atherton explained: 'A 1,000MW offshore wind farm would have revenue of around £560 million a year at current levels of subsidy and power prices. But only around £60 million of this revenue would get recycled to the local economy through operating the asset. The other £500 million would go to pay interest on the debt and provide a return for the shareholders. If the wind farm was built by a foreign company then these monies would leave Scotland.'

As noted earlier, the Dutch government has learned this lesson. They have recently reduced their support for renewables, cutting their total annual subsidy from 4.5 billion to 1.5 billion and vetoing any new onshore developments. This comes after they paid 4.5 billion to a German developer to install two 300MW offshore wind farms off the Dutch coast. Dutch politicians have woken up to something which so far hasn't occurred to Alex Salmond: the heavily subsidised green jobs which our taxpayers have to fund are actually being created in the countries that manufacture the turbines and not in the countries that install them.

It was interesting to note that the world's biggest wind turbine will shortly be erected at Methil in Fife, at the mouth of the River Leven. This 180-metre monster will be installed a few hundred metres offshore. It will be taller than London's Gherkin tower with a diameter greater than the London Eye and will be visible for miles around. But this massive feat of engineering will be undertaken by Samsung in South Korea and shipped to Scotland. Jobs will be created in Korea, paid for by Scottish electricity consumers.

But not all Korean energy companies are as keen on Scotland's renewable policy. The Scottish Government's wind industry plans received a major setback in April 2012 when Korean company Doosan Power Systems announced that it was scrapping plans for a £170 million research centre and turbine factory in Glasgow that would have created 1,700 jobs. Doosan's shock announcement, only 12 months after the deal had been unveiled with great fanfare by Alex Salmond, was blamed on 'sapping market confidence putting a question mark over the future development of the offshore wind market', according to industry media outlet Recharge. Doosan themselves put out a press statement saying only: 'In light of the overall economic conditions and liquidity issues in Europe, Doosan Power Systems decided in December 2011 to withdraw from its plans for developing offshore wind turbines in Scotland.'

Doosan's worries were echoed by another large offshore services company, Petrofac, who warned that only 30% of anticipated wind farm projects in the North Sea will ever be built because of problems raising the huge amount of money involved.

Petrofac's CEO, Ayman Asfari, said he was keen to build a wind business, but the economic conditions made it difficult. 'We think that between 30% and 50% of the targets set for Round Three of offshore wind expansion are likely to be realised. There are few utilities that have the financial strength to meet the required scale of necessary investment.'

Doosan and Petrofac's gloomy outlook followed hot on the heels of a statement by Magued Eldaief, Managing Director UK of General Electric Energy, who said that his company's proposed £100 million investment in wind turbine manufacture was being put on hold until UK ministers clarified their decisions on the future reform of the energy market. He was referring to remarks by Chancellor George Osborne that the UK would become uncompetitive if it continued to pursue environmental goals set by the EU.

Few people realise that UK wind power operators receive a higher subsidy per MWh than in other EU member states, including even Denmark and Germany. Subsidies in Ireland, Spain and Portugal are significantly lower than in the UK, even though their average generation costs are similar. Surely Scottish taxpayers and consumers should have the right to ask whether their money would not be better spent on increasing energy efficiency and investment in technologies that will actually keep our lights on? Instead, the race for wind is underway and soon, when the subsidies dry up, we may have to face a reality check.

When the bubble bursts?

With the true facts about wind power increasingly reaching the public domain, more and more Scots are becoming disillusioned with a policy of support for wind energy. Scotland's green energy myth is being exposed. The ranks of anti-wind groups are now swelling and as correctly pointed out by Ben Acheson in an article for ThinkScotland, 'an influential anti-wind lobby grows daily. A small group of nimbyists has evolved into a fully fledged, well-informed lobbying machine with scientists, academics, celebrities and even environmentalists backing small-scale community activists and national anti-wind groups such as CATS, Country Guardian and NOW (National Opposition to Windfarms)' (Acheson 24/08/12).

Their disillusion with turbines can be reinforced by evidence from abroad. Anyone driving across the desert from Los Angeles to Palm Springs or visiting the southernmost tip of Hawaii's stunningly beautiful Big Island will be met with a breathtaking sight – scores of rusting and broken-down wind turbines. These relics of what became the American 'wind rush' of the mid 1980s lie derelict and idle, a sad monument to the expensive folly of wind power. They were supposed to save the environment, but their abandoned hulks are now spoiling it. This abject lesson from the US should have resonated with the wind fanatics in the Scottish Government, but sadly it seems to have been ignored. Scotland's wind rush is underway and soon, when the subsidies dry up and economists realise the futility of investing in wind and turn towards another technology, Scotland's hills, glens and coasts will be scarred by similar giant, rusting carcasses.

The US lesson is one that we should carefully note. In the late 1970s when the oil crisis began to bite, the US Government decided it had to look elsewhere for sources of power. Some so-called experts persuaded influential politicians that wind could provide a serious alternative to fossil fuels. A construction programme began with giant industrial turbines being constructed at Kamaoa on Hawaii, Altamont (east of San Francisco), Tehachapi on the edge of the Mojave Desert and at San Gorgonio near Palm Springs. The whole exercise was supported by a generous sea of subsidies that provided rich returns for investors and landowners alike. The similarities with what is happening in Scotland today are uncanny.

In 1986 the price of oil tumbled and the subsidies began to dwindle. Suddenly investing in wind no longer made economic sense and when the big turbines began to break down, it was no longer financially viable to repair them, so they were simply abandoned and left to rot. Now few of the US wind farms still operate and those that do produce a negligible trickle of electricity. It is only now, decades later, that irate local authorities in states like California are threatening lawsuits against companies that have abandoned wind turbines and some are starting to be dismantled.

In Palm Springs, it took campaigners like the late Sonny Bono, former husband of Cher, to draw attention to these derelict monuments to the green energy myth. But in many cases where the company who had erected the turbines had gone bankrupt or disappeared, it was left to the hapless farmer or landowner to foot the bill for dismantling the steel monsters. Of an original 4,500 derelict turbines in California it is estimated that 500 are still standing. Scotland's farmers and landowners who have willingly allowed giant turbines to be built on their land should beware. Long after the subsidy tap has been turned off and the flow of cash into their pockets has dried up, they may find themselves responsible for dismantling the turbines and restoring the landscape to its former glory. This has certainly been the experience of their American cousins.

CHAPTER TEN

Alternative Energy Sources

When confronting the increasing antagonism to the current obsession with wind power, many people suggest that either solar, hydro, wave or tidal power might provide the answer. Such views are ill-founded. Hydropower in Scotland plays only a minor role, producing around 6% of Scotland's electricity, and currently there are no full-scale commercially operating wave and tidal power stations in the UK; solar energy, like wind power, can only form a tiny part of the overall energy mix. Biomass has none of the drawbacks of wind or solar power, but it is constrained by the need to grow food and provide raw materials for transport, biofuels and industrial processing.

Any sensible future energy mix may find a minor role for renewables. Solar thermal energy could be a solution to the domestic market and is efficient for water heating, but anyway its production is still expensive and discontinuous. Many of these problems could be overcome by new technological development, such as storage (fuel cells and others). It is important to debate whether energy from waste is renewable or not, but either way it should be part of an energy mix.

But it is essential that any future policy which seeks to enhance the security of energy supply must involve a varied and flexible energy mix. All of our eggs shouldn't be placed in the renewables basket! That advice has been ignored in Scotland with the SNP Government's obsession for wind. There are many actual and potential alternative sources of energy, such as biofuel plants, photovoltaic cells, hot rocks or geothermal plants, fuel cells (fixed and mobile), hydrogen production by CO_2-free energy sources, all of which are discussed in this chapter. Certainly energy-saving measures, including better insulation and microprocessors in buildings, will reduce electricity consumption, even out peaks and troughs, and increase efficiency.

There are other ways to cut radically our dependence on fossil fuels and in particular on oil. We need to double the efficiency of the oil and gas we use. We need to increase dramatically the energy efficiency of our homes, buildings and factories throughout the EU. We need to re-design a new generation of super lightweight cars and trucks, using advanced composites like carbon fibre, back-stopped by lightweight steels. We have the technology to produce lightweight vehicles which would not only double their efficiency compared to current cars and trucks, but would also improve safety.

We can apply the same technology to aircraft and heavy trucks, ultimately saving up to 52% of current oil usage. By turning to biofuels and other alternatives to oil, until such time as we can switch entirely to using carbon-free fuels like hydrogen, we could replace a further 20% of EU oil requirements, at the same time giving a huge boost to our beleaguered farming sector. However, we must not allow a race to biofuels to cause environmental destruction and to swap oil dependency for food dependency in Europe. Nor must we allow the Scottish Government's obsession with wind energy to plunge Scotland into a fuel crisis that could lead to blackouts in the near future.

Wave and tidal energy

While offshore wind farms may not be cost effective, many people think that wave or tidal energy could be exploited around Scotland's long coastline. A new marine energy park has been opened in the Pentland Firth to assist with the evaluation and development of future wave and tidal energy technologies. Once again, ambitious targets are being mooted with claims by the UK Government that energy from this stretch of turbulent water between Caithness and Orkney could produce a staggering 27 GW of electricity by 2050, enough to power around 20 million homes. However, although the European Marine Energy Centre (EMEC) has so far received £30 million in public funds from both the UK and Scottish Governments, this is quite literally a drop in the ocean. Developing this technology will require billions.

There are also other problems and pitfalls. Although tidal energy is more constant and more reliable than wind power, it is nevertheless intermittent. The tides only run for 16 hours daily, so tidal generators would still require base-load backup from fossil-fuel gas or coal-fired power plants. Wave power may have more potential, but could also be intermittent. In addition, wave surge converters currently under development are located in water around 10 metres deep, so like tidal generators, their construction could do unlimited harm to marine ecosystems like seagrass meadows and kelp forests that are themselves major blue carbon storage systems. So the jury is still out on wave and tidal energy. The vast level of investment necessary to bring them to a viable level of electricity generation, together with the potential for marine ecosystem damage, may render this technology obsolete before it gets a chance to develop.

Biofuels and impacts on biodiversity

Despite increasing public awareness of the need to protect and preserve biodiversity, scientists believe that between 150 and 200 species are being lost every 24 hours. Much of those losses can be attributed to climate change. We need to teach the public that biodiversity is valuable; it has an economic, social, aesthetic and practical value from which every one of us individually benefits. Biodiversity services purify the air we breathe, act as a global air-conditioning system, provide us with rainfall and oxygen and fertilise plants. We have never put a price tag on these ecosystem services because they are invaluable. But sadly, some people think that anything that is free has no value and therefore can be exploited and abused.

We have to take care that the policies we pursue are sustainable. For example, the drive to produce biofuels is causing global deforestation, which as well as releasing massive quantities of CO_2 into the atmosphere, could also lead directly to global famine. We are potentially creating a bigger global problem than we set out to resolve. In the US, vast quantities of maize are being converted to bio-ethanol. This in turn has led to huge tracts of the Amazonian rainforest being burned to make way for growing maize and soya as food crops to make up the shortfall.

Meanwhile the Indonesian rainforest is being torn up to make way for biofuel crops like palm oil to supply the EU market. Such policies are thus destroying the world's air-conditioning system while at the same time releasing millions of tonnes of CO_2 into the atmosphere. Deforestation is responsible for more greenhouse gas than all the world's cars, trucks, planes and boats combined.

Greed instead of care for the environment has become the defining feature of our strategy for tackling climate change and the race to biofuels is potentially threatening the lives of millions of people as the global population soars from its present six billion to an estimated nine billion by 2050. An extra six million people are born every month. By 2030 the world population will have expanded by such an extent that we will require a 50% increase in food production to meet anticipated demand. By 2080 global food production would need to double. But the reality is that an area the size of Ukraine is being taken out of agricultural food production every year due to drought and as a direct consequence of climate change. Global food production is declining rather than expanding and our headlong rush to produce biofuels is taking even more land out of food production.

We need to pay more attention to the ecosystem services provided by different aspects of our EU environment such as the peat bogs of Scotland, Ireland and Finland and the great cork oak forests of Portugal, Spain and France, as well as the seagrass meadows, kelp forests and maerl beds which surround the UK. All of these fragile ecosystems act as carbon sumps, absorbing and storing carbon for centuries. They form an essential part of our global air-conditioning system and we cannot allow them to deteriorate or to be destroyed by the race to develop renewables.

Energy efficiency

There is little doubt that we could save 75% of the energy we currently use by being more efficient. The Renewable Energy Foundation advocates improving energy efficiency because a large proportion of the UK housing stock is 100 or more years old. In fact, 21% of our houses were built before 1918, from which there is a huge heat loss. But even today it is shocking that we still allow homes to be built in Scotland with single-glazed windows and no loft insulation. Indeed in some listed buildings we *insist* on single glazing. Floor insulation, top-up loft insulation, low-energy lighting and A-rated condensing boilers would undoubtedly help, but triple glazing and proper insulation would cut our energy bills dramatically.

Energy conservation is priority number one and is simply about eliminating the need for energy in the first place through the reduction or elimination of unnecessary energy use. It subscribes to the notion that a kilowatt saved is more valuable than a kilowatt supplied. It is often achieved through behavioural change, such as switching appliances and devices off when they are not being used. But engineering solutions also have an important supporting role to play, for example through the provision of smart technologies to help consumers make conservation choices.

We need to improve the energy efficiency of both the demand side and the supply side. Enormous savings can be made by the use of more efficient domestic appliances, more efficient vehicles and more efficient heat delivery systems. The Institution of Mechanical Engineers suggests that the government should focus on incentivising efficiency savings in end-use electricity consumption. To do so, they could strengthen obligations on suppliers through the Carbon Emissions Reduction Target (CERT) scheme and phase out inefficient appliances such as plasma TVs. Smart meters for all households would also help conserve electricity by reducing consumption.

On the supply side, take a modern fossil-fuelled power station for example. The steam turbine generator system can have its efficiency significantly increased by the use of supercritical or ultra-supercritical steam, which gives a much larger power output from the same quantity of fuel. If the waste heat energy produced is harnessed and utilised, as is commonplace in most other European countries, the overall efficiency of the plant is greatly increased; the heat energy is provided from the same amount of fuel that would have been used to generate the electricity in the first place, thereby leading to an efficiency gain.

Only after we have properly addressed energy conservation and efficiency should we be thinking of renewables. Focusing on conservation and efficiency now would give us adequate time for appropriate research and development into renewable strategies so we can implement the right strategies for Scotland at the right time, instead of ploughing ahead with ill-conceived plans conceptualised to meet ill-conceived targets.

A nuclear future?

Mott MacDonald, one of the UK's leading energy consultants, who advise Whitehall, state in a report prepared for the Department of Energy and Climate Change that nuclear power is projected to be the least-cost option of the main generation types in the long term. At the moment, taking into account the first-of-a-kind (FOAK) costs, Mott MacDonald note that nuclear has a levelised cost of around £100 per MWh when costs of safe storage of waste and final decommissioning are included. In comparison, onshore wind is roughly comparable at £96 per MWh, and offshore wind is completely out of the park at an enormous £190 per MWh. Clean coal plants come in at around £145 per MWh and new gas turbines at £110 (DECC/Mott MacDonald 2010).

The 'UK Electricity Generation Costs' report noted that onshore wind is slightly cheaper than nuclear at present, but their figures are based on the costs of new 'third-generation' nuclear plant, which is still considered to be a less proven technology compared to onshore wind, a 'mature technology'. In light of this, Mott MacDonald articulate that 'in the longer term as nuclear moves to NOAK (nth-of-a-kind) status, and as carbon and fuel prices rise, nuclear is projected to become the least-cost main generation option with costs around £67 per MWh, some £35–45 per MWh below the least-cost fossil fuel options' (DECC/Mott MacDonald 2010).

Mott MacDonald correctly point out that 'determining the costs of generation is not an easy matter'. Comparing levelised costs is especially fraught with difficulties as the 'levelised cost of generation is the lifetime discounted cost of ownership of using a generation asset converted into an equivalent unit cost of generation in £/MWh or p/kWh. This is sometimes called a life-cycle cost, which emphasises the cradle-to-grave aspect of the definition' (DECC/Mott MacDonald 2010).

This is a reminder that, when comparing nuclear power with onshore or offshore wind energy, we must remember longevity in addition to efficiency and decommissioning costs. We must remember that nuclear power is the cheapest low-carbon generator, because virtually carbon-zero nuclear power has an 80% load factor, compared to wind turbines which have a load factor of only around 24–30% and require baseload backup to ensure the lights don't go out. New nuclear power plants have an estimated life of around 60 years, extendable to 120 years, while onshore and offshore wind turbines, or even tidal and wave systems, will require complete replacement every 20 years, covering much larger areas and using much larger quantities of concrete and steel than nuclear plants. Taking all this into account one would imagine that the average right-thinking person would conclude that new nuclear plant is a 'no-brainer'. But sadly not in Scotland, where our energy policy is designed by the SNP Government to achieve political rather than economic goals.

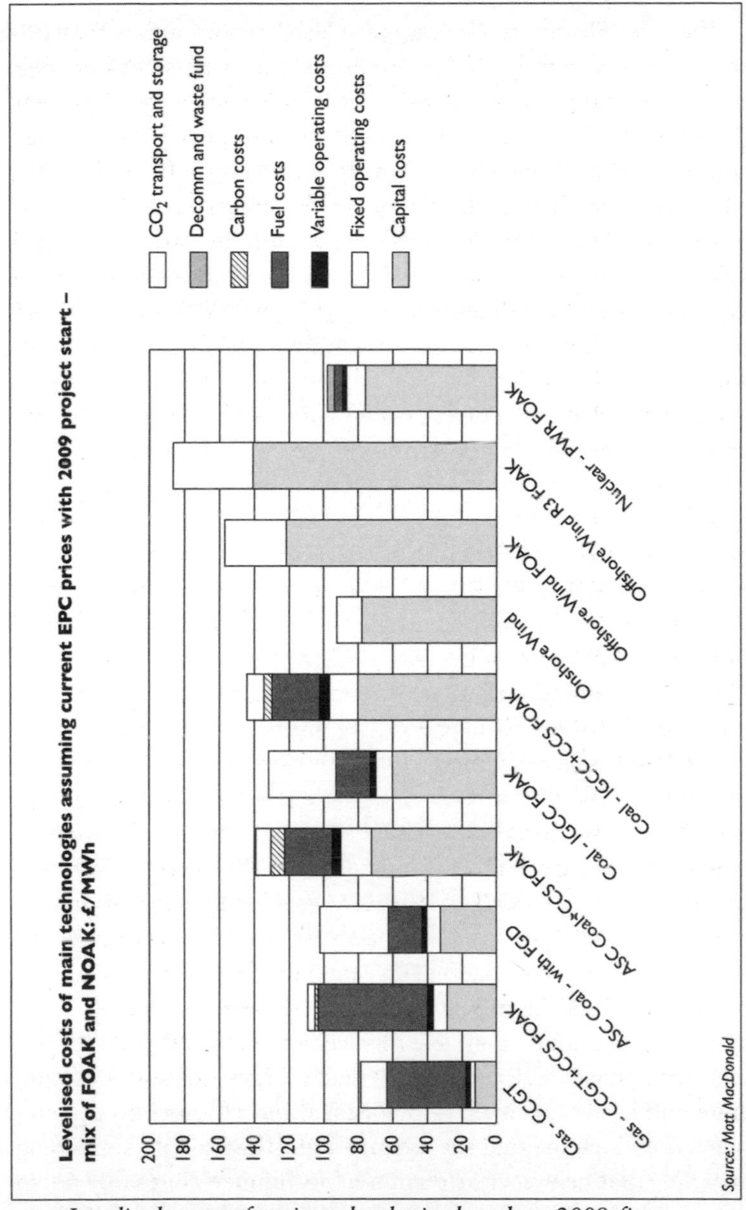

Levelised costs of main technologies based on 2009 figures (Source: DECC/Mott MacDonald 2010)

Nuclear power is virtually CO_2 emission free. Nuclear energy currently represents about 15% of the EU's total energy production, although this figure conceals wide differences between member states. As many as 15 of the 27 EU member states generate nuclear electricity. Countries like France rely almost entirely on nuclear generation and export large amounts of electricity to the UK via cross-Channel interconnectors. On the other hand, countries like Austria are entirely nuclear-free, while Germany has decided to phase out and close all of its nuclear plants by 2022.

In the wake of the Fukushima disaster of 2011 and public reaction to it, work is underway to develop nuclear reactors that won't melt down. There is still considerable interest in nuclear technology with 440 nuclear generating plants operating worldwide and 18 countries currently building or planning to build new nuclear reactors, from Turkey to Thailand – 62 nuclear plants are being built and a further 158 will be up and running by 2020. Leading the field in the nuclear race is China, with 13 current nuclear plants in operation and 27 under construction. Another 50 are at the planning stage. Six new nuclear plants are planned in America. So interest in nuclear technology is still buoyant.

However, the new nuclear fuels are what are really getting the scientists excited. In Canada, work is underway to coat uranium pellets with an oxide made from beryllium, a non-reactive metal. Uranium is a poor conductor of heat and the reactor core has to be raised to enormous temperatures to generate enough heat to produce steam. This is where the risk of meltdown can occur if the process is not handled properly.

In contrast, beryllium is a first-class conductor of heat, so the reactor core can be run at much lower temperatures without affecting the performance of the generator. This reduces the likelihood of a meltdown, although it doesn't rule it out completely. But the new, fourth-generation designs for nuclear reactors even deal with the problem of nuclear waste. The 'travelling wave' reactor and the PRISM reactor designed by GE Hitachi actually burn their own waste as a fuel source.

Despite the diverse attitudes to nuclear energy, its use guarantees a stable supply and long-term stability. The import of uranium as a primary fuel source from politically stable countries is more secure than any other primary energy sources. The new generation of nuclear plant is also highly attractive, with faster construction times, simpler designs and enhanced safety features. With the use of nuclear energy we are saving 720 million tonnes of CO_2 emissions (equal to the emissions from 212 million cars in Europe). It is clear that nuclear power will have an expanding role to play as a sustainable and dependable energy source in the future.

Nuclear energy currently provides about half of Scotland's baseload electricity. With Hunterston A on the Ayrshire coast decommissioned and Hunterston B and Torness within years of the end of their (extended) active lives (Hunterston B is due to close in 2016), it is absolutely imperative that new stations at Hunterston C and Torness are commissioned without delay. It is sheer madness to turn our backs on nuclear power.

We must ignore the siren wails of the Greens who cite Fukushima as a reason to abandon nuclear technology in Europe. The EU does not suffer from 9.2 Richter-scale earthquakes and tsunamis. What we do suffer from is the constant scaremongering pedalled by the Greens and their fellow travellers who are the people behind the current renewable wind blitzkrieg invading our country with industrial structures of concrete and steel, all for a small, intermittent trickle of electricity at vast cost to the consumer.

The many hundreds of skilled workers at Hunterston and Torness have provided safe, CO_2-emission-free energy to the National Grid for generations. We owe it to them and their families and to future generations of Scots who want to live in an environmentally sustainable world to order a new nuclear generating plant now.

The UK Government has decided that in future, new nuclear plants will be financed by the private sector. This will mean transferring all running and operating costs to private companies, although the long-term and controversial issue of dealing with nuclear waste may still involve some government finance. The introduction of the Non-Fossil Fuel Obligation acts as an incentive for providing financial support to nuclear generators, although it is not clear that the private sector is currently enthusiastic about constructing new nuclear plants in England. In January 2008, the UK Government announced plans to encourage private operators to build eight new nuclear power stations in England and Wales, although the Welsh Assembly has now said that it opposes a replacement plant at Wyfla. Meanwhile the Scottish Government has said that it will refuse planning approval for any new nuclear power plants in Scotland.

Estimates vary as to the cost of new third-generation nuclear power plants. The UK Government says that a new plant should cost around £2.8 billion, but some industry leaders claim that the costs will be nearer £5 billion. There are currently 16 operating reactors in nine separate nuclear plants in the UK. The cost of building eight new nuclear plants in England alone could therefore top £40 billion, excluding the decommissioning costs for the old power stations and the expensive disposal of nuclear waste. The Institute for Public Policy Research (IPPR) reported in June 2012 that building 18GW of new nuclear-generating capacity in England with ten new reactors could create more than 32,000 jobs and boost exports. This was despite an earlier announcement in March 2012, when leading nuclear sector companies E.ON and RWE npower stated that they were no longer interested in plans to develop new nuclear plants in England.

It is currently the UK Government's responsibility to deal with nuclear waste in Britain, some of which has come from the production of nuclear weapons. Most high-level nuclear waste is presently stored at Sellafield in Cumbria. The government-appointed Committee on Radioactive Waste Management recommended in 2006 that all future high-level waste should be disposed of underground at depths of 200–1,000 metres, in appropriate geological formations. They stated that a purpose-built nuclear waste repository should be constructed for this reason and that the waste should then be permanently buried with no intention that it should ever be retrieved in the future.

It was generally accepted that the UK would have a single site for nuclear waste disposal and that this site would probably best be located at Sellafield. It was recognised that this recommendation would be controversial and that it might take some years to implement such a proposal. There is no doubt that the whole question of decommissioning nuclear reactors and the permanent safe disposal of high- and intermediate-level nuclear waste is both sensitive and hugely expensive.

Thorium

Perhaps an alternative to either fossil fuels or traditional nuclear power is thorium. The thorium fuel cycle relies on the use of the naturally abundant thorium, or 232Th to use its scientific name. The thorium fuel cycle claims several potential advantages over a uranium fuel cycle, including thorium's greater abundance, superior physical and nuclear properties, better resistance to nuclear weapons proliferation and reduced plutonium and actinide production.

Many nuclear experts argue that thorium should replace uranium because it doesn't yield nasty, weapons-grade waste and its residues only last for a few hundred years, rather than the tens of thousands of years associated with uranium. Thorium can also work in conventional water-cooled reactors, although when combined with alternative reactor designs such as 'liquid fluoride' or 'molten salt' reactors, it can be made even more efficient.

A thorium reactor is currently being built in India and Flibe Energy, an Alabama-based company, is also constructing a thorium reactor in the US. Indeed the US built a thorium molten-salt reactor at Oak Ridge National Laboratory in the 1960s, but President Nixon abandoned all further developments in favour of uranium, so that he could obtain more weapons-grade waste.

Now companies like Flibe argue that the next generation of

nuclear reactors should be based on thorium, which is much more abundant and widely distributed in the earth's crust than uranium. In addition, they say that their favoured type of thorium reactor, liquid-fluoride (LFTRs), can be smaller, cheaper to run and much safer than conventional nuclear plants. As a tremendous bonus, these fourth-generation nuclear reactors could even burn up existing stockpiles of high-level nuclear waste, using it as a fuel. Discussions are already underway with Sellafield about this concept.

The UK Government has begun to discuss the potential for thorium reactors and it may yet become part of a new energy development agreement signed in April 2012 between UK Energy Secretary Ed Davey and US Energy Secretary Steven Chu.

Shale Gas

A possible way out of the looming crisis may be through the exploitation of our massive reserves of shale gas, natural gas trapped within deep shale formations, which can be exploited through a process known as fracking, where a water-based solution is forced into fissures, releasing the gas. Although it is a highly contentious issue, exploiting UK shale gas reserves may help us reduce our dependency on imported gas from geopolitically sensitive areas such as Russia, Iran and the Middle East. With an estimated 200 trillion cubic feet of shale gas deposits discovered in Lancashire alone, enough to power Britain for 65 years, we could be looking at the biggest energy find since North Sea oil in the 1960s. But it is typical of the hysterical nature of the climate change debate in Britain that this massive find has been either entirely ignored or robustly attacked as anti-green.

Even James Lovelock, the famous scientist behind the Gaia theory who once predicted doom and gloom for our planet through climate change, now accepts that shale gas might provide the answer. He told *The Guardian* (15/06/2012) that the UK should be 'going mad for fracking'. He argues that while not perfect, it emits about half the CO_2 that burning coal produces.

'Gas is almost a give-away in the US at the moment,' he says, because they have gone for fracking in a big way. 'Let's be pragmatic and sensible and get Britain to switch everything to methane,' he told *The Guardian*. 'We should be going mad on it,' he concluded. Lovelock keeps a picture of a wind turbine on his office wall to remind him how 'ugly and useless they are'.

With conventional reserves of natural gas running low, prospectors are increasingly looking towards unconventional resources like shale gas. Conventional gas is considered to be gas that, over many millennia, has become trapped in porous rock chambers and then sealed by another impermeable rock on top. In contrast, unconventional gas is natural gas that is contained in porous rocks which generally have interconnected spaces and boreholes which allow the gas to flow freely through the ground. So-called 'unconventionals' are often difficult to access because they are found in rocks of low permeability. Besides shale gas, other 'unconventionals' include tight gas (found in sandstone), gas hydrates (methane trapped in permafrost and ice caps) and coal-bed methane (found in coal seams).

In the UK, several areas have already been identified as having large potential shale gas reserves. The Department of Energy and Climate Change has estimated that there are 135 billion cubic metres (bcm) alone in the Bowland shale near Blackpool. Although they have not yet been explored, the British Geological Survey suggests that UK offshore reserves of shale gas could be five to ten times the size of onshore. Some observers estimate that UK offshore reserves are in excess of one thousand trillion cubic feet, which would put the UK in the top 20 countries with shale gas reserves worldwide.

Exploiting UK reserves of shale gas would take some time and more research needs to be conducted into the safety risks and costs, but there is potential. The government must recognise that shale gas will be a major part of the energy mix for the foreseeable future and they must therefore at least consider shale gas technology. But sadly, in Scotland, this seems not to be the case.

Hydraulic Fracturing (fracking) is the most commonly used

method of well stimulation during shale gas extraction. Following the initial drilling, fracking involves pumping tens of thousands of litres of 'fracking fluid' – water mixed with salts, soap and citric acid – into the well under high pressure. Cuadrilla, a leading shale gas producer, estimates that up to 13 million litres of water is often used during one fracking process. The mixture causes rock formations to fracture and release stored gases. But there is mounting suspicion that a diverse coalition of anti-fracking, pro-wind interests are not being entirely discouraged by the Scottish Government, who realise that the development of a buoyant shale gas industry in Scotland would reduce gas prices dramatically and stop the renewable energy sector in its tracks.

The exploitation of shale gas reserves has had a considerable impact around the globe. In Europe, Poland and France have particularly large reserves. Poland's shale gas production may even end its dependency on Russian gas, but France, which has decided to rely on its nuclear power plants, has outlawed fracking on spurious environmental grounds.

In America, such has been the success of the shale gas industry that gas prices have plummeted. The US used to import over 100 bcm of liquefied natural gas (LNG) every year. In 2011, this fell to a total of only 20 bcm. Prices for American consumers have dropped by 30%. But of course the process has its critics.

WWF Scotland said it has concerns around the process, including the contamination of water supplies by the 'fracking' fluids and from gas leaking into water supplies, creating risks of explosions. Such claims need closer examination. Boreholes for shale gas extraction commonly are drilled down to 2,000 metres or more underground, thousands of metres below the aquifer. The risk of water contamination is therefore negligible. Similarly, the opponents of shale gas often point to news reports of methane leaking through the water supply so that in some cases in America, people have actually been able to set fire to water coming from bathroom taps. However, this phenomenon was first observed in 1932, decades before shale gas was even known about. It is a natural occurrence in certain parts of the USA, where methane gas has saturated the rock strata and entered the aquifer. This has nothing whatsoever to do with fracking, but provides nevertheless a convenient photo opportunity for the 'FRACK OFF!' brigade.

Critics also claim that fracking caused a 1.5 Richter-scale earthquake during exploratory drilling near Blackpool in 2011. They point to the fact that in 2012, France's National Assembly voted to ban fracking and they urge the Scottish Government to follow the French example. Several environmental groups have expressed concerns about the sheer volume of water used during the process and the impacts that wells have on the landscape, although this is minimal compared to wind turbines.

But shale gas producers in America, like Royal Dutch Shell, claim that they know how to avoid these risks. They say that so long as well shafts are properly sealed with concrete, there is a negligible risk that fracking will contaminate groundwater. A 2011 study of shale gas wells in America by Duke University found no traces of fracking chemicals in groundwater. However, it concluded that faulty cementing on wells where there had not been full cementing below the level of the lowest groundwater source did lead to the migration of fracking fluid into the environment.

Furthermore, by eliminating venting and flaring, methane emissions can be kept to an absolute minimum and the risk of earthquakes, which has long been present in conventional oil, gas and even coal extraction, can be monitored and controlled. The International Energy Agency (IEA) says that implementing such precautions will add 7% to the cost of shale gas, but the industry believes this is an acceptable price to pay for a healthy and rapidly expanding sector.

In terms of the volume of water used during fracking, the Institution of Gas Engineers and Managers commented how fracking does not rank particularly high when compared to water-intensive industries such as paper and pulp manufacturers who use up to 130 million litres per year. Furthermore, the

Environment Agency has pointed out that much of the water used in shale gas extraction is recycled and reused, although this can concentrate the contaminants in the water.

And as for those who argue that burning shale gas is the antithesis of going 'green', the industry points out that it produces only half as much CO_2 as coal. Burning shale gas will therefore lead to a dramatic fall in emissions. In fact, in the US, emissions have fallen by 450 million tonnes in the past five years, more than any other country in the world, while, ironically, emissions in the EU have actually risen over the same period due to the increased use of coal-fired energy, because of the rising price of natural gas.

So the shale-gas revolution has not only shamed the wind industry by showing how to cut carbon emission for real, but it has also blown away the last vestiges of credibility in the argument that says supplies of fossil fuels will soon disappear, leaving no alternative but renewables, no matter what the cost. Even if the price of oil remains above $100 a barrel, it looks like supplies of cheap and plentiful gas are here to stay for many decades to come.

Scottish consumers are facing a double whammy at the hands of the renewable energy lobby; we are witnessing a relentless rise in our electricity bills to meet the cost of unreliable and heavily subsidised wind turbines, while at the same time gas prices are rocketing to meet increasing demand caused by the need for back-up energy sources for the 75% of the time that wind turbines are standing idle. The Institution of Gas Engineers and Managers has stated that Britain's shale gas reserves are unlikely to be great enough to be a 'game changer', but they would undoubtedly contribute to gas security and provide a lower-carbon alternative to conventional fossil-fuel generation. Despite this, in Scotland, our chance to exploit potentially enormous shale gas supplies, which could cause gas bills to plummet and create thousands of new jobs, is being quietly stifled by a chorus of protests and a government obsessed with wind.

The Hydrogen Economy

We should also be investing much more into developing the new sunrise technologies such as the hydrogen economy. So far, no one has invented an efficient way to store electricity. But hydrogen, which is the lightest and most abundant chemical element in the universe, can be readily stored and can provide an effective energy source. In Germany they are building hydrogen-powered cars, trains and ferries. Hydrogen-powered homes are under construction. We need to cut our dependency on fossil fuels and look to the future.

We are on the threshold of the third industrial revolution, which, like the first two, will be driven by the convergence of new energy regimes with new communication regimes. The first hydraulic agricultural societies – Mesopotamia, Egypt, China, and India – invented writing to manage the cultivation, storage and distribution of grain, which was used to feed slaves who, in turn, provided the energy to build the pyramids and cities and run the ancient economies.

In the modern era, the link between coal-powered steam technology and the printing press gave birth to the first industrial revolution. It would have been impossible to organise the dramatic increase in the pace, speed, flow, density and connectivity of economic activity made possible by the coal-fired steam engine using the ancient hieroglyphs and oral forms of communication.

This in turn was followed in the late nineteenth and first two thirds of the twentieth centuries by first-generation forms of electrical communication – the telegraph, telephone, radio, television, electric typewriters and calculators. These inventions, many of them Scottish, converged with the introduction of oil and the internal-combustion engine, enabling the organisation and marketing of the second industrial revolution.

Now the smart technologies which gave us the internet and vast distributed global communication networks will be used to reconfigure the world's power grids so that people can produce renewable energy and share it, people to people, in the same way they currently share information. The creation of a renewable energy regime, partly stored in the form of hydrogen and distributed via smart intergrids, will open the door to a third industrial revolution, transforming the way we live in the 21st century.

Renewable forms of energy – solar, wind, hydro, geothermal, ocean waves and biomass – make up the first of the three pillars of the third industrial revolution. Although only providing a tiny amount of the current global energy mix, these sunrise technologies are attracting massive amounts of private and public investment as governments around the world mandate targets and benchmarks for their widespread introduction.

To maximise renewable energy and minimise its cost it will be necessary to develop storage methods that facilitate the conversion of intermittent supplies of these energy sources into reliable assets. As we have seen throughout this narrative, renewable energy, by its very nature, is unreliable. When the wind isn't blowing, or the sun isn't shining, or the water isn't flowing because of drought, electricity can't be generated. But if we use some of the electricity generated while renewable energy is abundant to extract hydrogen from water, which can then be stored for later use, society will have a continuous supply.

Hydrogen is the lightest and most abundant element in the universe and when used as an energy source, the only by-products are pure water and heat. Our spaceships have been powered by high-tech hydrogen fuel cells for more than 30 years.

Electricity produced from renewable sources can be used to split water into its separate elements of hydrogen and oxygen by the process of electrolysis. Hydrogen can also be extracted directly from energy crops, animal and forestry waste and organic rubbish – so called biomass – without the need for the electrolysis process. By utilising hydrogen as a storage carrier for all forms of renewable energy, society can guarantee a constant supply of power. Of course the cost of hydrogen is still relatively high, but new technological breakthroughs and economies of scale are dramatically reducing costs year on year. Moreover, hydrogen-powered fuel cells are at least twice as efficient as the internal-combustion engine.

The EU has become the first superpower to flag up the transition to the third industrial revolution by making a binding commitment to producing 20% of community energy requirements from renewable sources by 2020. The European Commission has established a massive research initiative called the Hydrogen Technology Platform to further develop the hydrogen economy. Meanwhile Chancellor Angela Merkel of Germany has committed £350 million to hydrogen research in Germany and called for a third industrial revolution in several speeches she has made.

Already hundreds of hydrogen-powered fuel cell forklifts, scooters, cars, buses and trucks are being used throughout the EU. A hydrogen fuel cell submarine is in operation in Germany and hydrogen-powered ferries are under development in Germany and the Netherlands. Europe's first hydrogen-powered train was completed in 2010. BMW has stated that it will produce one million hydrogen-powered cars per year by 2020. These will be high-performance vehicles, capable of travelling at over 120mph and able to travel up to 500 miles on a single fuelling of hydrogen. They are already rolling out hydrogen filling stations across Germany. These cars will be entirely CO_2 free. The only thing that will emerge from their exhaust pipes will be drops of pure water.

At least the SNP Government has recognised the benefits of hydrogen, with Alex Salmond himself announcing Scotland's first ever hydrogen-powered fleet of buses (*Daily Record* 14/08/12). The First Minister said that £3.3 million had been given to Aberdeen City Council to finance the purchase of ten hydrogen buses which are expected on the city's streets by early 2014. Scottish and Southern Energy Power Distribution (SSEPD) will develop an integrated 'whole hydrogen system', which will use wind energy to produce sufficient electricity to electrolyse water, so that hydrogen can be produced and stored as a fuel source for the buses. The project, costing £1.65 million, has been part funded by Scottish Enterprise and the European Commission. Perhaps a semi-useful way of utilising wind power has been found at last.

But spending £3.3 million on a few hydrogen-powered buses is not a realistic response to this potentially ground-breaking new carbon-free technology. Compared to the billions being flung at the wholly unreliable wind sector, this is a sad example of just how wrong-headed the SNP Government's energy policy has become.

So the first two pillars of the third industrial revolution, a shift to renewable energy and an aggressive hydrogen fuel cell technology research and development programme – are now in place. The third pillar – the re-configuration of the EU power grid, along the lines of the internet, allowing businesses, industries, individuals and homeowners to produce and share their own energy with each other – is already being tested by power companies in Europe. By fostering a hydrogen infrastructure and a continent-wide intelligent intergrid, the EU can help create a sustainable economic development plan and turn the dream of an integrated single market into a reality for our 500 million citizens in the first half of the 21st century. Maybe then the Scottish Government will come to its senses. But by that time it might be too late.

While the world searches for an effective alternative to fossil fuels, some radical experiments are taking place, including research into high-powered laser energy, or HiPER. The HiPER consortium involves seven nations (USA, Japan, South Korea, Russia, China, Canada and the UK), two regional governments, and a great many international institutions and industries, in a long-term research project to develop laser-driven fusion energy. A preparatory three-year project was completed in 2011. A definition phase is now under way with full construction due to begin in 2014. It is hoped a working laser-driven fusion fuel plant should be in operation by 2020.

Fusion is the bonding together of small atoms, the output from which reaction releases a huge amount of energy that can be captured to drive a gigawatt-scale power plant. Most of the activity is taking place around the PETAL laser in the Aquitaine region of France, where lasers are used to compress a shell of deuterium and tritium fuel to a very high density. An immensely high-powered laser, roughly ten thousand times the power of the entire UK National Grid, is then focused into the dense DT fuel for a few million millionths of a second, acting like a match and raising it to fusion temperatures (~100 million °C).

Enthusiasts of the HiPER project claim that it will provide energy security, because its main fuel source will be seawater. They say that it will provide fuel on such a mammoth scale that it will meet all of the planet's requirements. Furthermore, there are no greenhouse gas emissions associated with the process and no long-lived radioactivity. What is clear is that the HiPER science programme will open up many new areas of research in a wide range of scientific disciplines including astrophysics in the laboratory and the behaviour of matter in truly extreme conditions.

Carbon Capture and Storage

Many environmental experts have long predicted that a system of capturing and storing carbon dioxide emissions from fossil-fuel burning power stations and from huge industrial processes like iron and cement works is the only realistic way to deal with the problem of climate change. They argue that to limit global warming to no more than +2°C – which is widely accepted as the top limit for safety – carbon emissions will have to be halved by 2050.

The International Energy Agency (IEA) says that to achieve carbon capture and storage (CCS) to this extent, we will have to construct 100 new capture facilities worldwide by 2020 and over 3,000 by 2050. The problem is, at the moment there are only eight such facilities and not a single one of them is attached to a power station. Others, mostly in North America, are under construction, but many such projects have been abandoned as the sector loses confidence in continuing government support for a technology that is so hideously expensive it cannot survive without massive subsidy.

CCS is expensive because it involves a complex procedure that runs gases which are to be processed through a solution of amines or ammonium carbonate. These react with CO_2 to form soluble chemicals called carbamates and bicarbonates which can be treated to release their load of CO_2, which can then be piped off and stored. The remaining exhaust gases arising from this process, mostly nitrogen, are safely vented into the atmosphere, while the liberated amines or ammonium carbonates can be recycled for reuse.

The problem with all of this is that the process as currently developed is itself a huge consumer of energy. It would use about one quarter of the energy output from an average coal-fired power station, which, according to Howard Herzog – a chemical engineer at the Massachusetts Institute of Technology – would cost between $50 and $100 per tonne of CO_2 stored. Such costs are just not sustainable, which is why many of the CCS projects under construction have stalled or been ditched entirely, including a demonstration plant at Longannet in Fife. The Department of Energy and Climate Change announced its intention not to proceed with the Longannet CCS project in October 2011 due to unsustainable costs. So unless some technological breakthrough occurs soon, slashing the cost of CCS, it looks like this is a non-starter for tackling climate change.

Algae

If you have ever tried to wash away the green, slimy stuff that grows up walls and on paving stones, you will know how resilient it is and how quickly it can grow back. This green stuff is algae and it is currently being examined as a potential energy source for the future. These tiny green plants produce energy-storing molecules called lipids from which oil can be extracted for fuel. During the Beijing Olympics in 2008, the world's media were greatly absorbed by the sight of hundreds of Chinese soldiers fighting to clear hundreds of tonnes of algae growing daily across the canals earmarked for rowing events. It was clear that these tiny green organisms can multiply at a terrific rate, almost overwhelming attempts to control their abundant growth.

It also was clear from this Chinese spectacle that algae can grow even in the most hostile environments, such as brackish or even salt water, thriving even on raw sewage and carbon dioxide, which it draws out of the atmosphere.

Scientists have established that the oil-bearing properties of algae are impressive. While corn only yields around 70 litres of oil per acre harvested, rapeseed around 480 litres per acre and palm oil 2,400 litres per acre, algae can yield an astonishing 55,000 litres of oil per acre. However, there is a problem. The big oil-producing algal blooms grow slowly. The fast-growing algal blooms produce very little oil. So scientists are now working on genetically modified algae that can be tricked into growing fast, like the algae at the Beijing Olympics, while also producing huge amounts of oil. If they crack this conundrum, they could develop an almost perfect energy source. A lot of investment is now going into the development of algae as a new fuel source, particularly in China and India. But even the Americans say that they are on the way to producing algafuel at only $2 per gallon and that they will even have jet aircraft quality algafuel in production by 2013 at only $3 per gallon.

Photovoltaic (PV)

The sale of solar panels peaked in Scotland in 2011, before subsidy cuts imposed by the UK Government caused the photovoltaic sector to falter. On 1 August 2012 Feed-in Tariff rates for solar photovoltaic (PV) plants smaller than 4 kW fell to £0.16/kWh and contract terms fell to 20 years from 25 years. The UK's Solar Trade Association (STA) attempted to put a brave face on it, stating that investing in PV will continue to be attractive for homeowners and businesses despite these cuts.

Solar panel electricity systems, also known as solar photovoltaics (PV), capture the sun's energy using photovoltaic cells. These cells don't need direct sunlight to work – they can still generate some electricity on a cloudy day. The cells convert the sunlight into electricity, which can be used to run household appliances and lighting. But they are expensive, produce only a trickle of electricity and, like other parts of the renewable energy industry, cannot survive without subsidies.

Paying hefty subsidies to solar panel installers and generous FiTs to people who installed them in their homes and businesses seemed like a good idea during the economic good times, but it now looks like an unaffordable extravagance during the age of austerity. As governments cut the subsidies, PV companies file for bankruptcy. Three solar energy companies in Germany – Solon, Solar Millennium and Solarhybrid – have filed for bankruptcy since December 2011.

Now Jean-Louis Bal, head of France's Renewable Energy Association, has said that the photovoltaic industry currently has production capacity twice that of market demand, meaning that a lot of manufacturers, including even some big Chinese companies, are facing severe financial difficulties.

There will always be a role for photovoltaic panels in a diverse energy mix, but it will inevitably be a minor role. Perhaps combining PV panels with another emerging low-carbon technology like hydrogen could provide a way forward. For example, solar panels mounted on the roof of a house, even in far-from-sunny Scotland, could produce enough electricity to conduct electrolysis on a supply of water to produce hydrogen. Hydrogen fuel cells could then provide all the energy needs of the house, including cooking, heating and lighting, virtually free of charge and with zero carbon emissions.

As noted in this chapter, hydrogen can also be used as a fuel for powering internal-combustion engines by combustion or electric motors via hydrogen fuel cells.

Hydro

Hydro power currently produces about 6% of Scotland's electricity and, according to the experts, there is some potential for new hydro schemes. Hydro was one of the first ever clean and green renewables, being utilised successfully in Scotland decades before the terms 'clean' and 'green' were even recognised.

More than half of Scotland's 145 hydroelectric schemes are in the Highlands and Islands area, with most modern plants achieving energy conversion rates topping 90%. So their efficiency is three times greater than onshore or offshore wind and comparable to nuclear power. Total hydro generation capacity in Scotland is about 1,500 MW.

There are a number of new projects in the planning stage including two large-scale new pumped storage schemes in the Great Glen area of the central Highlands proposed by Scottish and Southern Energy (SSE). The two plants, with a combined generation capacity of 900 MW, will be built at Core Glas, north-west of Loch Lochy, near Fort Augustus, and Balmacaan, near Invermoriston. They will be the first pump storage systems to be built in Britain in more than three decades. They operate by pumping water into an upper reservoir or loch during low power demand periods when electricity prices are low. The water is then released down to a lower reservoir when demand rises and prices are high, generating electricity through a hydro-power plant.

SSE already operates a 300MW pumped storage scheme at Foyers, on the south side of Loch Ness. In 2008, the company completed Britain's first large-scale conventional (non-pump storage) hydro electric station for more than 50 years: the £150 million, 100 MW Glendoe plant, near Fort Augustus.

However, according to Stuart Young of Stuart Young Consulting, the entire pumped storage hydro capacity in the UK can provide up to 2,788 MW for only five hours, then it drops to 1,060 MW, and finally runs out of water after 22 hours. It is clear therefore that more pumped storage hydro capacity can certainly be built, but the massive investment required would have little impact on climate change, delaying global warming by a matter of hours at vast cost.

There is also growing interest in rolling out micro-hydro schemes across Scotland. These small-scale hydro-electric schemes would attract FiTs and could contribute to local community development needs. A report to the Scottish Government, published in January 2010, showed a substantial range of viably developable small hydro schemes with a combined potential capacity of 1,204 MW, sufficient to supply around one million homes. The study described considerable untapped potential amongst more than 7,000 possible schemes, almost all of them smaller than 5MW capacity. So far, however, micro-hydro projects have met with a range of obstacles from planning and licensing authorities, including SEPA, the Scottish Environmental Protection Agency.

Geothermal

Geothermal energy is the natural heat of the Earth stored in rocks and the fluids within them. Most of this energy is derived from the decay of radioactive materials within the Earth. As a result of this heat production, there is a flow of heat towards the Earth's surface, resulting in temperatures that increase with depth (the geothermal gradient). Geothermal systems which heat water are widely used for providing a supply of hot water and central heating in homes and businesses throughout the UK. Although the initial installation costs are high, geothermal energy provides a long-term saving on rising electricity bills and as such is becoming increasingly popular.

Conclusion

There are many new and exciting developments in the field of energy and electricity generation. Fuel efficiency can save a great deal of the energy we currently waste, while new carbon-free technologies can provide a potentially cost-effective and economically viable future. The clear conclusion must be that any stable energy strategy must involve a diverse mix of technologies in which renewables will inevitably have a role to play. As shown in the foregoing section on hydrogen, solar or wind power could be an efficient way of producing electricity to utilise for hydrolysis – splitting hydrogen and oxygen from a supply of water, so that the hydrogen can be stored and used as a carbon-free power source. What is equally clear is that the obsessive pursuit of renewables to the virtual exclusion of all other energy sources is an impossible dream which will have devastating consequences for the Scottish economy.

CHAPTER ELEVEN

Sensible Environmentalism

Protecting the environment *is* a noble and essential cause. But as I have shown over the course of this book, using wind energy in Scotland as an example, we are not doing it properly. Every day we hear more and more lurid claims of melting ice caps, vanishing polar bears, thawing glaciers and extremes of weather. But environmental hysteria doesn't help anyone. We need to take a long, hard, sober look at the facts and decide on the most practicable way to deal with the problems we identify. Knee-jerk responses to global environmental concerns will never lead to a satisfactory outcome.

It is also important to note that nowadays nobody deliberately tries to destroy the environment; nobody sets out to change the climate and end the world. But the combination of environmental hysteria and romanticised 'I am saving the planet' Green groupthink has got us to where we are today. The same politicians who fell for the global hysteria surrounding the 'millennium bug', are now being driven towards knee-jerk solutions to problems that require a more rational and carefully considered approach. Nowhere is this more evident than in Scotland.

These politicians have allowed romanticism to overshadow fact-based, scientific evidence. They trot out the global-warming arguments, merely regurgitating what they have heard from others, without actually knowing the facts. Every storm, flood, and drought is blamed on climate change. Sage heads nod in agreement. Something, anything, must be done to avert disaster.

Thus groupthink takes hold and suddenly targets are set and politicians unite in their determination to save the planet. It is in exactly this kind of milieu that bad decisions are taken. I believe this is what has happened in Scotland and to a lesser extent across the UK as a whole.

In 1986, a New York environmentalist called Jay Westervelt coined the term *greenwashing* to describe the practice of hotels sticking notices on bathroom walls inviting guests to 'save the environment' by not sending their towels to be laundered every day. Westervelt noted that far from trying to save the planet or be environmentally friendly, these hotels were mainly obsessed with increasing their own profits by minimising laundry costs. Westervelt decided to label this and other dubious 'environmentally friendly' acts, such as putting an image of a forest on a bottle containing harmful chemicals, as greenwashing.

I am increasingly convinced that the mad dash for wind energy has fallen foul of the greenwashing syndrome. Proponents of industrial turbines proclaim that they are the greatest thing since sliced bread. We commonly see illustrations of wind turbines adorning packages of supposedly environmentally friendly products. Banners bearing pictures of wind turbines are dotted throughout the corridors of the European Parliament in Brussels and adorn the walls of major European airports, purporting to illustrate Europe's successful campaign to combat global warming.

Like the ubiquitous hotel bathroom signs, wind turbines have now become associated with everything that is clean, green and environmentally friendly. But like the hotel bathroom signs, this international example of greenwashing is nothing more than a clever disguise for an exercise in blatant profiteering, which achieves little or nothing for the environment, but fills the pockets of power companies and landowners alike, at the direct cost of the consumers, while ticking the green boxes of politicos.

We hear constantly that Scotland has 25% of all the wind resources of Europe and that this 'free' resource will be permanently available and utilisable for the benefit of every man, woman and child. No one stops to ask how on earth anyone can have accurately measured the constant wind speeds across the entire European continent and decided that one quarter of the total wind blows across Scotland. It is simply typical of the romanticised nonsense trotted out by the pro-wind lobby.

The combination of greenwashing and groupthink inevitably leads down a political cul-de-sac, where more and more outlandish targets are set and ever greater commitments are made, so that politicians can wave their environmental credentials like a flag. I am certain that this is what has happened in Scotland. Having trumpeted their anti-nuclear and pro-renewables position from the opposition benches at Holyrood, the SNP suddenly found themselves with an actual majority in the Scottish Parliament. By now convinced by their own rhetoric, they clamoured to outdo Westminster, Brussels, and indeed the rest of the world, by setting preposterous targets to produce the equivalent of 100% of electricity generation equivalent from renewables by 2020, regardless of the cost to the consumers and the destruction of Scotland's landscape. This strategy has won praise for Alex Salmond from the likes of Al Gore, but it is the opposite of sensible environmentalism.

Every sensible person knows that cost is important. If you can't afford it, then it doesn't matter how good it is; it simply is not worth it. That goes for environmental policies too. Why should we simply ignore how much it costs, just because it might help the environment by an insignificant amount? Indeed, the worst environmental problem of all in the world today is poverty and making 900,000 people in Scotland fuel poor, due directly to an ill-considered energy policy, is exacerbating rather than tackling the issue. The transfer of money from the poor to the rich in pursuit of meaningless environmental targets is not something in which we should take pride.

In Scotland, the most simple way for the majority of urban dwellers to demonstrate their support for the battle against climate change is to cheer on renewable energy and decry opponents of wind turbines as 'climate change deniers' or 'nimbyists', safe in the knowledge that the constant 'whump, whump' and flicker effect of a giant industrial turbine will never besmirch their view or disturb their sleep in the wealthy suburbs of Glasgow,

Edinburgh or Aberdeen. The SNP jumped on the environmental bandwagon years ago and as it gained speed, they found it more and more difficult to jump off. It is now racing out of control towards disaster.

Of course what we need is sensible environmentalism. Of course we need to protect the environment. But we need a practical, straightforward approach. Everyone needs to take a step back and realise that the world is not about to implode. The climate may well be changing but instead of hysteria-driven alarmism and scaremongering, we need to deal with the problem sensibly. We need to work together to come up with solutions that work. A solution that works in Brazil may not work in Scotland. But we may have ideas that could be adapted elsewhere.

We don't have to go back to living in caves. We don't have to go back to square one. But we need to get back on a sensible track. Green groupthink has taken over. We need to return to science and logic. We need to get rid of the view that the environment has to be returned to the state it was in 2,000 years ago, before any sort of real civilisation existed. That simply isn't feasible. We just need to do it sensibly.

Sensible environmentalism is what Dr Patrick Moore is all about. Dr Moore was one of the co-founders of Greenpeace, but he resigned in the mid 1980s, increasingly concerned that it had been hijacked by political activists. Now he says he believes in sustainable development, rather than the politically ideological environmental goals of the green movement. Moore asserts that the green lobby now commonly uses fear and lies to promote their message. In January 2012, Dr Moore described wind power as 'a destroyer of wealth and negative to the economy'.

Patrick Moore's book *Confessions of a Greenpeace Dropout: The Making of a Sensible Environmentalist* describes how he became sensible as Greenpeace became increasingly senseless, adopting an anti-science, anti-business and 'downright anti-human' agenda.

Moore's idea of sensible environmentalism, he says, 'requires embracing humans as a positive element in evolution rather than viewing us as some kind of mistake'. Moore says we should be growing more trees, using more wood and building dams for hydroelectric energy. He also believes that 'nuclear energy is essential for our future energy supply, especially if we wish to reduce our reliance on fossil fuels. It has proven to be clean, safe, reliable, and cost-effective.'

Moore supports geothermal heat pumps, genetic engineering to improve crop yields, and aquaculture, 'including salmon and shrimp farming'. He says in his book that this will be 'one of our most important future sources of healthy food' and 'will take pressure off depleted wild stocks and 'employ millions of people productively'.

Dr Patrick Moore's version of sensible environmentalism, coming from one of the founders of Greenpeace, is worthy of careful consideration. He clearly illustrates the pitfalls of making the wrong choice; of going in the wrong direction. Decision-makers and opinion-shapers need a global one-stop shop where they can examine all of the options and decide what is sensible and what is senseless.

Such a one-stop shop was in fact endorsed in the final communiqué from the Rio+20 summit. The Astana 'Green Bridge' Partnership, floated in Rio by Kazakhstan, seeks to create a single entity, a think tank or a one-stop shop, where governments and energy companies could go for advice on the best systems for their specific circumstances, where the most sensible environmental options could be examined in detail. They promoted the concept that this facility could be staffed by leading experts from around the world who could examine the precise weather patterns, economic circumstances and system options for any specific country or even region and come up with the best possible advice for cost-effective and carbon-efficient energy projects. Advice like this would help avoid the mad dash for wind that has blighted the UK.

The United Nations' Economic and Social Commission for Asia and the Pacific (ESCAP) has endorsed the Astana 'Green Bridge' Initiative with a view to creating the establishment of a

Europe-Asia-Pacific partnership that will outline the blueprints for a shift from the current conventional development patterns to green growth. The core idea is aimed at getting world experts to work together on research into emerging and cutting-edge policy solutions and tools, looking at ways that these solutions for green growth can be implemented and financed in the most cost-effective manner. Kazakhstan has confirmed that they will meet the initial costs for getting the project off the ground, including head-hunting and paying handsome salaries to a leading pool of global climate-change experts and economists. Kazakhstan's interesting 'Green Bridge' Partnership could prove to be a major building block towards achieving an environmentally sustainable and sensible future for all of us, hopefully before the lights go out in Scotland as a result of the SNP's obsession with wind power.

I have tried to demonstrate in this book how the use of massive subsidies has driven a misguided and ill-thought-out energy policy with potentially devastating consequences for the UK and Scotland in particular. The penultimate chapter explored the plethora of options, quite apart from the many and varied types of renewable energy, that are currently available for producing electricity or capturing and storing carbon in a 'green' economy. But it is this confusing range of options that has frequently caused politicians and decision-makers to head off in the wrong direction, leaving a trail of destruction in their wake.

I am sure that renewables will have a role to play in any sensible future energy strategy. For example, the utilisation of wind-, solar- or hydro-generated electricity to electrolyse a supply of water so that hydrogen can be produced, stored and used as a CO_2-free energy source is something worth aiming for. In the meantime, energy efficiency, cutting the amount of energy we waste, is an essential way forward to reduce electricity bills and carbon emissions simultaneously. So too is the need for a continued reliance on nuclear power as a carbon-free and reliable energy source which has served us well for generations.

The key lesson that I hope readers will take from this book is that an obsessive race to produce 100% of our equivalent electricity demand from renewables by 2020 in Scotland is just not sensible, not economically viable and not environmentally sustainable. Recognising that the situation in Scotland is reaching tipping point, *Spectator* magazine decided to hold its first-ever debate in Edinburgh, entitled 'Scotland's energy policy is just hot air' at the National Museum of Scotland on 19 September 2012.

On their arrival, audience members were asked to cast a vote on whether they supported the proposition, opposed it, or whether they had yet to make their minds up. Before the start of the debate, 66 were for the proposition, 36 against and 74 sat on the fence. After a two-hour debate chaired by Andrew Neil and featuring speeches from myself, Andrew Montford, Prof. Stuart Haszeldine and Niall Stuart (CEO of Scottish Renewables), the audience was asked again if they agreed or disagreed with the proposition that 'Scotland's energy policy is just hot air'. Tellingly, 126 voted for, 50 were against and nobody was undecided.

Perhaps the worm turned in Edinburgh on 19 September 2012. Perhaps confronted by the arguments, a majority of Scots may yet rebuff Scotland's green energy myths and reject this government's turbine tyranny.

Bibliography

Achard, F.; Eva, H.D.; Stibig, H.J.; Ayaux, P.; Gallego, J.; Richards, T. and Malingreau, J.P., (2002) 'Determination of Deforestation Rates of the World's Humid Tropical Forests', *Science* 297: 999–1002.

Acheson, B.W. (24/08/2012), 'All at Sea: offshore wind farms will leave Scotland feeling blue'. Available at: http://www.thinkscotland.org/thinkpolitics/articles.html?read_full=11588&article=www.thinkscotland.org (Accessed on 15/09/2012).

Acheson, B.W. (14/09/2012), 'The Winds of Change are Blowing: So What Next?'. Available at: http://www.thinkscotland.org/thinkpolitics/articles.html?read_full=11627&article=www.thinkscotland.org (Accessed on 15/09/2012).

Acheson, B.W. (1/10/12), 'The Hunt for Green October'. Available at: http://www.thinkscotland.org/thinkpolitics/articles.html?read full=11676&article=www.thinkscotland.org (Accessed on: 30/12/2012).

Acheson, B.W. (1/10/12), 'Forgotten communities are left high and dry by wind farms'. Available at: http://www.thinkscotland.org/thinkpolitics/articles.html?read full=11718&article=www.thinkscotland.org (Accessed on: 30/12/2012).

Allison, E.H.; Perry, A.L., Badjeck, M.C.; Adger, W.N.; Brown, K.; Conway, D.; Halls, A.S.; Pilling, G.M.; Reynolds, J.D.; Andrew, N.L. and Dulvy, N.K. (2009), 'Vulnerability of national economies to the impacts of climate change on fisheries', *Fish and Fisheries*.

Andrews, J.E.; Samways, G. and Shimmield, G.B. (2008), 'Historical storage budgets of organic carbon, nutrient and contaminant elements in saltmarsh sediments: Biogeochemical context for managed realignment, Humber Estuary, UK', *Science of The Total Environment* 405: 1–13

Angelsen, A. (Ed.) (2008), 'Moving ahead with REDD: Issues, options and implications', CIFOR (Bogor, Indonesia).

Arrigo, K.R. (2005) 'Marine micro-organisms and global nutrient cycles', *Nature*, 437:7057 p349, doi:10.1038/nature04159.

BIS (2012), 'An international comparison of energy and climate change policies impacting energy intensive industries in selected countries'. Available at: http://www.paper.org.uk/documents/12-527-international-policies-impacting-energy-intensive-industries.pdf (Accessed on 15/09/2012).

Boorman, L. and Hazelden, J. (1995), 'Salt marsh creation and management for coastal defence' in: Healy, M.G.; Doody, J.P. (1995), *Directions in European Coastal Management* pp175–183.

Borum, J.; Duarte, C.M.; Krause-Jensen, D. and Greve, T.M. (2004), 'European seagrasses: an introduction to monitoring and management', The M&MS Project, Copenhagen.

Bouillon, S.; Borges, A.V.; Castañeda-Moya, E.; Diele, K.; Dittmar, T.; Duke, N.C.; Kristensen, E.; Lee, S.Y.; March, C. and Middelburg, J.J., Rivera-Monroy, V.H.; Smith III, T.J. and Twilley, R.R. (2008), 'Mangrove production and carbon sinks: A revision of global budget estimates', *Global Biogeochemical Cycles* 22: GB2013, doi:10.1029/2007GB003052.

Bryden, D.M; Westbrook, S.R.; Burns, B.; Taylor, W.A. and Anderson, S. (2010), 'Assessing the economic impacts of nature-based tourism in Scotland', Scottish Natural Heritage Commissioned Report No. 398.

Cambridge Economic Policy Associates (CEPA) (2011), 'Note on impacts of the CfD-FIT support package on costs and availability of capital and on existing discounts in power purchase agreements', Report for the Department of Energy and Climate Change, Cambridge: CEPA, June 2011.

CBC News (1/10/2011), 'Ontario wind power bringing down property values', Available at: http://www.cbc.ca/news/canada/ottawa/story/2011/09/30/ontario-wind-power-property-values.html (Accessed on 15/09/2012).

Consumer Focus Scotland (2012), 'Consumer Focus Scotland's response to the Economy, Energy and Tourism Committee's inquiry into the Scottish Government's renewable energy targets', www.consumerfocus-scotland.org.uk

Daily Mail (24/03/2012), 'Anger as third of council land is earmarked for wind farms'.

Daily Record (14/08/12), 'Aberdeen could be home to first fleet of hydrogen buses after £3.3million investment is revealed', Available at: http://www.dailyrecord.co.uk/news/science-technology/hydrogen-buses-could-be-operating-in-less-1260754 (Accessed on 15/09/2012).

DECC/Mott MacDonald (2010), 'UK Electricity Generation Costs

Update', Report for the Department of Energy and Climate Change, Brighton: Mott MacDonald, June 2010.
Department of Energy and Climate Change (2011), 'Annual Report on Fuel Poverty Statistics 2011', London: DECC.
Department of Energy and Climate Change (2011), 'UK climate change sustainable development indicator: 2010 greenhouse gas emissions, provisional figures', London: DECC.
Department of Energy and Climate Change (DECC) (2011), 'Planning our electric future: a white paper for secure, affordable and low-carbon electricity', London: DECC, CM 8099, July 2011.
Der Spiegel (08/16/2012), 'Grid instability has industry scrambling for solutions'.
EEA (2009), 'Europe's onshore and offshore wind energy potential: An assessment of environmental and economic constraints', EEA Technical Report No. 6/2009, EEA: Copenhagen.
Energy Action Scotland (2007), 'Special Report on Fuel Poverty'. Available at: http://www.theclaymoreproject.com/uploads/associate/365/file/EAS%20Publications/Election_Special_2007.pdf (Accessed: 15/09/12).
Energy Action Scotland (2011), 'UK Fuel Poverty Monitor', Glasgow, UK.
Energy Action Scotland (2012), 'UK Fuel Poverty Monitor – Fuel Poverty: The State of the Nations 2011'. Available at: http://www.theclaymoreproject.com/uploads/associate/365/file/EAS%20Publications/Monitor%202011%20Final%20_2_.pdf (Accessed: 15/09/12).
E.ON Netz (2005), 'Wind Report 2005', Bayreuth: E.ON Netz Gmbh.
European Energy Review (27/08/2012), 'We are aiming for a transformation – a re-industrialisation along the lines of a green economy'. Available at: http://www.europeanenergyreview.eu/site/pagina.php?email=ellie.jones@segec.org.uk&id_mailing=303&toegang=11b9842e0a271ff252c1903e7132c-d68&id=3826 (Accessed on 15/09/12).
European Union, 'Directive 2009/28/EC of the European Parliament and of the Council', (Renewable Energy Directive), 23 April 2009.
European Union (1985), 'Environmental Impact Assessment Directive (85/337/EEC)'. Available at: http://ec.europa.eu/environment/eia/eia-legalcontext.htm (Accessed on 15/09/12).
European Wind Energy Association, 'Wind in power: 2011 European Statistics'.
Forestry Commission (2012), 'Forestry Commission Great Britain/

England Annual Report and Accounts 2011–12'. Available at: http://www.official-documents.gov.uk/document/hc1213/hc00/0055/0055.pdf (Accessed on 15/09/2012).

Glasgow Caledonian University, Moffat Centre and Cogentsi (2008), 'The economic impacts of wind farms on Scottish tourism', Scottish Government, Edinburgh.

Good Practice Wind Project (2012), 'Involving energy and environmental issues in school curricula or extracurricular activities to provide a foundation for balanced decision-making later in life'. Available at: http://www.project-gpwind.eu/index.php?option=-com_content&view=article&id=175&Itemid=337#read-the-good-practices-about-this-specific-recommendation (Accessed on 17/09/12).

Gray, L. (2012), 'Hay Festival 2012: Simon Jenkins: "I cannot believe we allow wind farms to be built in areas of natural beauty"', *Telegraph*. Available at: http://www.telegraph.co.uk/culture/hay-festival/9322637/Hay-Festival-2012-Simon-Jenkins-I-cannot-believe-we-allow-wind-farms-to-be-built-in-areas-of-natural-beauty.html (Accessed on 15/09/2012).

The Guardian (07/02/2012), 'KPMG refuses to publish controversial green energy report'. Available at: http://www.guardian.co.uk/environment/2012/feb/07/kpmg-green-energy (Accessed on 15/09/2012).

The Guardian (27/02/2012), 'Wind power still gets lower public subsidies than fossil fuel tax breaks'. Available at: http://www.guardian.co.uk/environment/2012/feb/27/wind-power-subsidy-fossil-fuels (Accessed on 15/09/2012).

The Guardian (15/06/2012), 'James Lovelock: The UK should be going mad for fracking'. Available at: http://www.guardian.co.uk/environment/2012/jun/15/james-lovelock-interview-gaia-theory (Accessed on 15/09/2012).

Hanning, C. (2009), 'Sleep disturbance and wind turbine noise'. Available at: http://docs.wind-watch.org/Hanning-sleep-disturbance-wind-turbine-noise.pdf (Accessed on 15/09/2012).

Heck, Jr., K.; Carruthers, T.; Duarte, C.M.; Hughes, R.A.; Kendrick, G.; Orth, R. and Williams, S., 2008, 'Trophic transfers from seagrass meadows subsidize diverse marine and terrestrial consumers', *Ecosystems* 11: 1198–1210.

Hendriks, I.E.; Sintes, T.; Bouma, T. and Duarte, C.M., 2007, 'Experimental assessment and modelling evaluation of the effects of seagrass (*P. oceanica*) on flow and particle trapping', *Marine Ecology Progress Series* 356: 163–173.

The Herald (22/08/2011), 'Aristocrats cash in with subsidised wind farms'. [online] Available at: http://www.heraldscotland.com/news/home-news/aristocrats-cash-in-with-subsidised-wind-farms.14814956 (Accessed on 15/09/2012).

The Herald, (03/02/2012), 'University in wind farm row'.

The Herald (30/06/2012), 'Wind farm cash pledges "are bribery"', David Ross: Highland Correspondent.

The Herald (31/07/2012), 'Wind turbines may halve bat activity'.

Hughes, G. (2011), 'Why is Wind Power So Expensive?', The Global Warming Policy Foundation GWPF Report 7.

Huhne, C. (2011), 'The economics of climate change', Speech to Corporate Leaders' Group, London, 29 June 2011.

Iberdrola (2012), 'Flying Cloud 44 MW Wind Power Plant'. Available at: http://www.iberdrolarenewables.us/pdf/Flying%20-Cloud%20Fact%20Sheet.pdf (Accessed on 15/09/12).

IEA (2011), Speech by Nobuo Tanaka at World Energy Summit.

Institute of Acoustics (2012), Discussion Document on 'A good practice guide to the application of ETSU-R-97 for wind turbine assessment'.

Institution of Mechanical Engineers (2009), 'Energy and Buildings', Policy Document. Available at: http://www.imeche.org/Libraries/Position_Statements-Energy/EnergyandBuildingsIMechEPolicy.sflb.ashx (Accessed on 15/09/12).

Institution of Mechanical Engineers (2009), 'The Energy Hierarchy', Policy Document. Available at: http://www.imeche.org/Libraries/Position_Statements-Energy/EnergyHierarchyIMechEPolicy.sflb.ashx (Accessed on 15/09/12).

Institution of Mechanical Engineers (2009), 'Sustainable Electricity Supply', Policy Document. Available at: http://www.imeche.org/Libraries/Position_Statements-Energy/ElectricitySupplyIMechE-Policy.sflb.ashx (Accessed on 15/09/12).

Institution of Mechanical Engineers (2010), 'Carbon Capture and Storage', Policy Document. Available at: http://www.imeche.org/Libraries/Position_Statements-Energy/Carboncaptureandstorage-forgasplants.sflb.ashx (Accessed on 15/09/12).

Institution of Mechanical Engineers (2010), 'UK Offshore Wind Round 3', Policy Document. Available at: http://www.imeche.org/Libraries/Position_Statements-Energy/UK-Offshorewind-Round-3.sflb.ashx (Accessed on 15/09/12).

Institution of Mechanical Engineers (2011), 'UK Electricity Generation: Cost-Effective Management', Policy Document. Available at: http://www.imeche.org/Libraries/Public_Affairs/IMechE_Electricity_Generation_PS_Feb_2012.sflb.ashx (Accessed on 15/09/12).

Institution of Mechanical Engineers (2011), 'UK Energy Security Power', Policy Document. Available at: http://www.imeche.org/Libraries/Position_Statements-Energy/Energy_Security_Policy_Statement.sflb.ashx (Accessed on 15/09/12).

Institution of Mechanical Engineers (2011), 'Scottish Energy 2020?', Policy Document. Available at: http://www.imeche.org/Libraries/2011_Press_Releases/IMechE_Scottish_Energy_Report.sflb.ashx (Accessed on 15/09/12).

Institution of Mechanical Engineers (2012), 'Electricity Storage', Policy Document. Available at: http://www.imeche.org/Libraries/Public_Affairs/IMechE_Electricity_Storage_v7.sflb.ashx (Accessed on 15/09/12).

Joskow, P.L. (2010), 'Comparing the costs of intermittent and dispatchable electricity generating technologies', Working Paper 10-013, Center for Energy and Environmental Policy Research, MIT.

KPMG (2011), 'Rethinking the Unaffordable; Understanding the true cost of Green Transition', Economic Policy Centre.

Lea, R. (2012), 'Electricity Costs: The folly of wind power', Civitas: London.

Macdonald, A. (2012), *Wind Farm Visualisation: Perspective or Perception*, Whittles Publishing, Dunbeath, Scotland.

Mackay, J. (2012), *Scotland's Beauty at Risk*, Culross the Printers, Coupar Angus.

Marine Scotland (2011), 'Blue Seas – Green Energy: A Sectoral Marine Plan for Offshore Wind Energy in Scottish Territorial Waters – PART A: The Plan', Scottish Government, Edinburgh.

Martínez, M.L.; Intralawan, A.; Vázquez, G.; Pérez-Maqueo, O.; Sutton, P. and Landgrage, R., 2007, 'The cost of our world: ecological, economic and social importance', *Ecological Economics* 63: 254–272.

Minnesota Department of Health, Environmental Health Division (2009), 'Public Health Impacts of Wind Turbines'. Available at: http://www.health.state.mn.us/divs/eh/hazardous/topics/windturbines.pdf (Accessed on 15/09/2012).

Moore, P. (2010), *Confessions of a Greenpeace Dropout: The Making of a Sensible Environmentalist*, Beatty Street Publishing Inc., USA.

Mountaineering Council of Scotland (MCofS) (15/06/2012), 'MCofS Launches Manifesto on Onshore Wind Farms'. Available at: http://www.mcofs.org.uk/assets/manifesto%20on%20onshore%20wind%20farms.pdf (Accessed on 15/09/2012).

Nellemann, C.; Corcoran, E.; Duarte, C.M.; Valdés, L.; De Young,

C.; Fonseca, L.; Grimsditch, G. (Eds), 2009, *Blue Carbon* – A Rapid Response Assessment, United Nations Environment Programme, GRID-Arendal, www.grida.no

Nellemann, C.; Hain, S. and Alder, J. (eds) (2008), 'In Dead Water: Merging of climate change with pollution, over-harvest and infestation in the world's fishing grounds', UNEP Rapid Response Assessment, GRID Arendal, Norway.

OECD (2002), 'OECD Workshop on Environmentally Harmful Subsidies', Paris, France.

Press and Journal (25/07/09), 'RSPB Scotland joins opposition to £800m Shetland wind farm plan'. Available at: http://www.pressandjournal.co.uk/Article.aspx/1323472?UserKey= (Accessed on 17/09/12).

Raven, J. A. and Falkowski, P. G. (1999), 'Oceanic sinks for atmospheric CO_2', *Plant Cell Environ* 22, 741–755.

RYA (2004), 'Sharing the Wind: Recreational Boating in the Offshore Wind Farm Strategic Areas', The Royal Yachting Association and the Cruising Association.

The Scotsman (03/09/2012), 'Offshore wind farm plan for over 300 turbines'.

Scottish Daily Mail (20/08/2012), 'Buddhist monks flee "health risk" wind farm scheme', by Graham Grant, Home Affairs Editor.

Scottish Government (04/04/2012), 'Viking wind farm approved'. Available at: http://www.scotland.gov.uk/News/Releases/2012/04/viking04042012 (Accessed on 17/09/2012).

Scottish Government (20/09/2012), 'Draft Budget 2013–2014'. Available at: http://www.scotland.gov.uk/Publications/2012/09/7829 (Accessed on 22/09/2012).

Scottish Government Social Research (2010), 'The Economic Impact of Wildlife on Tourism in Scotland', Queens Printers of Scotland, Edinburgh.

Scottish Natural Heritage (2012), 'Climate Change and Nature in Scotland'. Available at: http://www.snh.org.uk/pdfs/publications/corporate/Climatechangenaturescotland.pdf (Accessed on 15/09/2012).

SEI (2009), 'The Need for Sound Carbon Accounting in Scotland', Policy Brief, Stockholm Environment Institute.

Sharman, H.; Leyland, B. and Livermore, M. (2011), 'Renewable Energy: Vision or Mirage?', The Scientific Alliance, www.adamsmith.org.

SQW Consulting (2009), 'Scottish Golf Tourism Market Analysis', Reports to Scottish Enterprise.

Stevenson, S.J.S. (2011). 'The Rape of Britain: Wind Farms and the Destruction of Our Environment', Bretwalda Policy Papers No. 5.

Strathspey & Badenoch Herald (25/10/2006), 'Pylon plans could jeopardise Highlands' future on screen', Gavin Musgrove.

Sunday Post (19/08/2012), 'Monks Flee Wind Farm', Euan Duguid.

Sunday Post (26/08/2012), 'Wind farms didn't blow in a blizzard of apprentices', Iain Harrison.

Sunday Telegraph (17/06/2012), 'Subsidies for onshore wind farms "to be axed by 2020"'.

Sunday Times (08/04/2012), 'Turbines are scarring Scotland's wilderness'.

Sunday Times (22/07/2012), 'Council Tax Cuts For Homes Near Wind Farms'. Available at: http://www.wind-watch.org/news/2012/07/22/council-tax-cut-for-homes-near-wind-farms/ (Accessed: 17/09/12).

Sunday Times (29/07/2012), 'No wind farms in wilderness, say most Scots'.

The Telegraph (21/08/2011), 'The aristocrats cashing in on Britain's wind farm subsidies'. Available at: http://www.telegraph.co.uk/earth/energy/windpower/8713128/The-aristocrats-cashing-in-on-Britains-wind-farm-subsidies.html (Accessed on 15/09/12).

The Telegraph (30/06/2012), 'Wind farm pylons: steel giants of the glens'.

The Telegraph (22/08/2012), 'SNP proposes wind farm "propaganda" for the classroom'. Available at: http://www.telegraph.co.uk/news/uknews/scotland/9490886/SNP-proposes-wind-farm-propaganda-for-the-classroom.html (Accessed on 17/09/12).

The Times (18/01/2012), 'Millions blown paying wind farms to close', Tim Webb.

The Times (13/07/2012), 'Green costs "may drive factories out of UK"', Tim Webb.

Tittensor, R. (2010), *From Peat Bog to Consider Forest: An Oral History of Whitelee, its Community and Landscape*, Packard Publishing Limited, Chichester, UK.

Travers, J. (2010), *Green & Gold: Ireland: A Clean Energy World Leader?,* The Collins Press, Cork.

UKCES (2011), 'Maximising employment and skills in the offshore wind supply chain, Volume 2 – supply chain case studies', Evidence Report 34, August 2011.

UKERC (2010), 'Great Expectations: The cost of offshore wind in UK waters – understanding the past and projecting the future'.

UK Government, 'The Renewable Energy Review', Committee on Climate Change, May 2011, London.

UNEP (2008), 'Reforming Energy Subsidies: Opportunities to Contribute to the Climate Change Agenda', United Nations Environment Programme Division of Technology, Industry and Economics.

University of Oxford (2012), 'Towards a low carbon pathway for the UK'.

University of Stirling (31/07/12), 'Eco-friendly microturbines need to be bat-friendly say Stirling researchers'. Available at: http://www.stir.ac.uk/2012/eco-friendly-microturbines-need-to-be-bat-friendly-say-stirling-researchers/name-30074-en.html (Accessed on 15/09/2012).

VisitScotland (2007), 'The Tourism Prospectus'. Available at: http://www.visitscotlandannualreview.com/content/pdfs/tour_prosp (Accessed on 15/09/2012).

VisitScotland (2009), 'Scottish Golf Tourism Market Analysis: Report to Scottish Enterprise'. Available at: http://www.visitscotland.org/pdf/SQWGolf%20Market%20Analysis%20Report.pdf (Accessed on 15/09/2012).

VisitScotland (2012), 'Wind Farm Consumer Research'. Available at: http://www.visitscotland.org/pdf/Insights%20Wind%20-Farm%20Topic%20Paper.pdf (Accessed on 15/09/2012).

Waycott, M. et al., *Accelerating loss of seagrasses across the globe threatens coastal ecosystems,* Proceedings of the National Academy of Sciences 106 (30) 2009.

Wilson, T.; Robertson, J. and Hawkins, L. (2012), 'Fuel Poverty Evidence Review: Defining, Measuring and Analysing Fuel Poverty in Scotland', Scottish House Condition Survey and Research Team, http://www.shcs.gov.uk

World Wind Energy Association (2011), 'World Wind Energy Report 2010', WWEA, Bonn, Germany.

Young, S. (2012), 'Gas, oil and coal prices were subsidised by £3.6bn in 2010 – or were they? An investigation on behalf of Communities Against Turbines Scotland', Stuart Young Consulting Ltd.

Index